DISCARDED
P9-AQW-699

Dilemmas in Animal Welfare

Dilemmas in Animal Welfare

Edited by

Michael C. Appleby
World Society for the Protection of Animals (WSPA), UK

Daniel M. Weary
University of British Columbia, Canada

and

Peter Sandøe
University of Copenhagen, Denmark

CABI is a trading name of CAB International

CABI
Nosworthy Way
Wallingford
Oxfordshire OX10 8DE
UK

Tel: +44 (0)1491 832111
Fax: +44 (0)1491 833508
E-mail: info@cabi.org
Website: www.cabi.org

CABI
38 Chauncey Street
Suite 1002
Boston, MA 02111
USA

T: +1 800 552 3083 (toll free)
T: +1 (0)617 395 4051
E-mail: cabi-nao@cabi.org

A catalogue record for this book is available from the British Library, London, UK

Library of Congress Cataloging-in-Publication Data

Dilemmas in animal welfare / edited by Michael C. Appleby, Daniel M. Weary and Peter Sandøe.
p. ; cm.
Includes bibliographical references and index.
ISBN 978-1-78064-216-1 (hbk)
1. Animal welfare. I. Appleby, Michael C., editor of compilation. II. Weary, Daniel M., editor of compilation. III. Sandøe, Peter, editor of compilation. IV. C.A.B. International, publisher.
[DNLM: 1. Animal Welfare. HV 4708]

HV4708.D54 2014
179'3--dc23

2013042143

ISBN-13: 978 1 78064 216 1

Commissioning editor: Sarah Hulbert / Julia Killick
Editorial assistant: Emma McCann
Production editor: Simon Hill

Typeset by AMA DataSet Ltd, UK.
Printed and bound in the UK by CPI Group (UK) Ltd, Croydon, CR0 4YY.

Contents

Contributors

Appleby, Michael C., World Society for the Protection of Animals, UK. E-mail: michaelappleby@wspa-international.org

Beausoleil, Ngaio J., Massey University, New Zealand. E-mail: N.J.Beausoleil@massey.ac.nz

Bennett, Pauleen, La Trobe University, Australia. E-mail: Pauleen.Bennett@latrobe.edu.au

Corr, Sandra, University of Nottingham, UK. E-mail: Sandra.Corr@nottingham.ac.uk

de Passillé, Anne Marie, University of British Columbia, Canada. E-mail: amdepassille@gmail.com

Edwards, Sandra A., University of Newcastle, UK. E-mail: sandra.edwards@ncl.ac.uk

Franco, Nuno H., Universidade do Porto, Portugal. E-mail: NFranco@ibmc.up.pt

Hötzel, Maria José, Universidade Federal de Santa Catarina, Brazil. E-mail: mjhotzel@cca.ufsc.br

Magalhães-Sant'Ana, Manuel, Universidade do Porto, Portugal. E-mail: mdsantana@gmail.com

Molento, Carla Forte Maiolino, Animal Welfare Laboratory, Federal University of Paraná, Brazil. E-mail: carlamolento@ufpr.br

Olsson, I. Anna S., Universidade do Porto, Portugal. E-mail: olsson@ibmc.up.pt

Palmer, Clare, Texas A&M University, USA. E-mail: cpalmer@philosophy.tamu.edu

Rushen, Jeff, University of British Columbia, Canada. E-mail: RushenJ@mail.ubc.ca
Sandøe, Peter, University of Copenhagen, Denmark. E-mail: pes@sund.ku.dk
Weary, Daniel M., University of British Columbia, Canada. E-mail: dan.weary@ubc.ca

Preface

An important approach to dilemmas is, of course, discussion among people who may have different viewpoints. It seemed obvious to the editors that this should be part of the preparation of this book. This proved time-consuming but worthwhile, so it illustrated the demanding nature but also the value of such communication:

- Initial ideas and outlines for chapters were circulated among authors for comment and refinement.
- A workshop was planned for July 2012 (in Vienna, Austria, prior to a conference). We requested draft chapters in good time, so that each could be read by other authors, one of whom then led discussion on that topic at the workshop. Those discussions proved lively and productive, recorded by a rapporteur so that everyone could use the ideas developed. We believe that they helped to clarify many of the issues, and the arguments needed to address them.
- A further round of revisions followed, in which authors were also encouraged to refer to other chapters for useful parallels and differences in the dilemmas considered.

One of the other interesting outcomes of the workshop was that some common themes emerged. For example, discussions about different categories of animals, such as domestic and wild animals in relation to conflicts between welfare and conservation, often assume that those categories are distinct. On the contrary, there is a continuum between fully domestic and fully wild animals, and between many other categories. Recognizing such common themes does not remove dilemmas, but may help to identify analogies and useful approaches from quite distinct fields. Some of these themes are addressed in the Introduction.

Introduction: Values, Dilemmas and Solutions

1

Michael C. Appleby,[1]* Daniel M. Weary[2] and Peter Sandøe[3]
[1]*World Society for the Protection of Animals, UK;* [2]*University of British Columbia, Canada;* [3]*University of Copenhagen, Denmark*

1.1 Introduction

Why a book on dilemmas in animal welfare? Thinking seriously about issues relating to animal welfare often means confronting a range of uncertainties over welfare outcomes, definitions of basic concepts, methodologies and appropriate decision making. Our hope is that confronting these uncertainties will improve our recognition of some of the gaps in our understanding and allow us to communicate our views in a more transparent way. We believe that such consideration is a real need for all those who confront the subject, including professionals working in the area of animal welfare (welfare scientists, philosophers, legislators, campaigners), people concerned for the welfare of animals under their care (pet owners, farmers, scientists using experimental animals) and members of the general public, for example those urged to support an animal charity or trying to decide what sort of eggs to buy.

Animal welfare provides a rich source of dilemmas, in part because opinions about the appropriate course of action are rooted in human values. The plural, 'values', is important here because individual humans have values that may be partially or completely in conflict (thus, a person may believe in both freedom of speech and the right of people not to hear speech that offends them), and different people may have different values that give rise to disagreements about decisions that affect animal welfare.

*E-mail: michaelappleby@wspa-international.org

1.2 What is a Dilemma Relating to Animal Welfare?

Imagine your family has a much-loved 4-year-old dog. One day you notice a growth on the side of her neck. You take her to the vet, who takes a biopsy and sends it off to be tested. The test result shows that the dog has a malignant cancer and is likely to die within 3 months if not treated. There is chemotherapy available that will probably allow the dog to live for another year but which also causes negative side effects, including appetite loss and fatigue. You are faced with a dilemma: euthanizing your loyal companion within the next few weeks or imposing an unpleasant procedure that may buy more time. In the face of such dilemmas, we are torn between competing concerns: neither choice is ideal and the harm imposed cannot be compared easily.

Different people may disagree about how to rank the competing values. Some members of your family may put more weight on the sanctity and duration of life and believe that the life of the dog should be prolonged at almost any cost; others may put more weight on minimizing suffering and will believe that prompt euthanasia is best for the dog's well-being. Cases may also involve uncertainty about different outcomes. For example, dogs vary in how well they respond to chemotherapy. So, there will be uncertainty about the outcome of the chemotherapy for the prognosis as well as the side effects.

This volume addresses a range of dilemmas. It deals with how, when deciding on the best way to treat an animal or groups of animals, one can be torn either between different concerns relating to the physical and affective welfare of the animal or between concern for the welfare of the animal and other, competing concerns. Different people may, as mentioned, disagree about how to resolve such dilemmas; these disagreements may relate to different views about how to handle uncertainties, for example about measuring and summing welfare concerns. Ultimately, they will typically be grounded in conflicting ethical perspectives.

1.3 Dilemmas within Animal Welfare

Some dilemmas can occur over how we conceptualize animal welfare: which aspects of animal welfare we value most and how we prioritize competing aspects. Thus, some cases pit concerns about animal health against those about affective states or the ability to engage in natural behaviours. For example, Edwards and Bennett's chapter on tail docking (Chapter 2) describes how concerns about the practice relating to pain and loss of natural function can be weighed against benefits to health. In this example and others, the facts associated with different cases may lead to different recommendations. Tail docking dairy cows was thought to improve udder health, but a series of studies has failed to show any health benefit from docking; this makes the decision to stop docking cows on an animal welfare basis an easy one. Other cases are more difficult. As Edwards and Bennett explain, pigs with intact tails are more likely to experience outbreaks of tail biting, leading to painful injuries and even death. In this case, one view would be that docking is the price that must (in current housing and management systems) be paid to avoid the risk of more serious injury. An alternative and conflicting view would be that it is wrong to cut off parts of animals to suit them to existing production systems and that the way forward must be to change the systems rather than mutilating the animals.

The idea of imposing a short-term pain or frustration for some longer-term health benefit also emerges in the chapter by Sandøe, Corr and Palmer (Chapter 3). In this case, the question is how we resolve the competing claims of fulfilling a dog's current wants (more food, more often) and keeping stress levels low, with reducing the incidence of diseases related to obesity. Rushen and de Passillé discuss (in Chapter 10) a related dilemma: keeping social animals in isolation (and in conflict with their motivations) with the hope of reducing the risk of injuries (due to social competition) and disease (due to direct contact between animals). In this case again, the resolution of different issues will depend on the empirical evidence that can be brought to bear and will vary with how the animals are housed and managed.

In all three examples, we can also see the competing values of harm minimization versus maximizing the opportunities for good welfare. Keeping sows in gestation crates prevents harmful competition, but also greatly limits the sows' ability to engage in normal locomotion and social behaviours that are probably important to a good life for the animals.

Although there are disagreements about the relative weighting of different aspects of animal welfare, most people working in the field agree that animal health, affective states and the ability to express motivated behaviours are all important. There is less consensus on where we should draw a line or gradient of moral concern within an issue. Weary's chapter (Chapter 11) challenges the reader to consider when certain types of negative affective states deserve special attention, by developing a new standard to identify when 'suffering' occurs. Cases where different sources of negative affect interact are likely to be a special concern, such as when animals repeatedly experience a painful procedure and develop a fear of the location or handlers associated with the procedure. If the fear can be avoided (e.g. by training the animals to approach the handler freely for a food reward), then the pain (e.g. from the injection) is of less concern.

1.4 Dilemmas within a Wider View of Animal Ethics

There are also disagreements about how other animal-centric arguments should be weighed against more traditional welfare values. A key example is the taking of a life. Most welfarists will argue that death per se is not a welfare concern, so long as the animal experiences a good death. Moreover, welfarists are typically concerned about individuals but struggle with how to make a trade-off between significant harm imposed on a few animals versus lesser harm imposed on many animals. Even if the sum of harm (however that is calculated) is smaller in the first case, it may still seem fairer to spread the harm more evenly among a larger group of animals. The chapter by Franco, Magalhães-Sant'Ana and Olsson (Chapter 4) takes us through a series of cases that discuss these conflicts, such as when it might be better to use more animals in an experiment if this means that each animal experiences less harm from the procedure. Palmer's chapter (Chapter 9) discusses how colonies of feral cats are 'managed' by sterilizing the animals, which results both in animals typically living longer and in there being fewer animals to deal with as fewer are born.

Palmer's chapter also challenges us to consider how welfare harm and welfare benefits vary for different types of animals. Sterilizing feral cats may help reduce colony size and reduce attacks on wild birds; keeping owned cats indoors will also reduce these attacks. In both cases, the suggestion is to impose one type of harm (e.g. preventing cats from engaging in a highly motivated natural behaviour) to avoid a painful death for other animals. The value we place on different types of animals also emerges in Beausoleil's chapter (Chapter 8), where we see the same species (the brushtail possum) treated with respect in Australia but exterminated (sometimes using painful and slow-acting chemicals) in New Zealand. The chapters by Molento (Chapter 7) and Palmer show us how the same species (dogs and cats, respectively) can sometimes be seen as a valued companion and at other times as a troublesome pest, leading to very different outcomes for the same blameless animals. Appleby's chapter (Chapter 6) explicitly takes on the issue of divergent values ascribed to different animals. He argues that we should preferentially use animals of high economic value in our farming systems as these animals are most likely to receive high levels of individual care.

Actions that favour the welfare of some animals may also come into conflict with values related to the ecosystem. Beausoleil discusses this conflict in her example of poisoning possums to save native birds, and Palmer discusses in her example the conflict of when to kill feral cats to save birds. Human health concerns can also come into conflict with concerns for animal welfare. Molento's chapter addresses how urban dog control programmes are often motivated by a desire to reduce the spread of diseases that dogs are thought to carry. Similarly, the chapter by Hötzel (Chapter 5) argues that decisions about livestock production systems 'must consider animal welfare, human health and nutrition, the environment and food security'.

1.5 Dealing with Dilemmas

There is no single, uniform way in which people can progress in the face of an animal welfare dilemma. One proven approach is to identify better ways of keeping or dealing with the animals in question that diminish or resolve the dilemma. The chapter by Beausoleil provides an example of using science to identify poisons that cause the least amount of pain and other negative experiences and thus reduce the conflict between the suffering of the target pest species (possums) and the environmental benefits that are thought to come from the death of these animals. In this way, applied animal welfare science often helps ameliorate welfare dilemmas.

But even when using the best choice of poison, the possums probably experience an unpleasant death, and definitely experience a shortened life, so in this case we are left with the conflict between animal welfare harm (to the possum) versus environmental benefits (to the endemic wildlife).

One common area of disagreement relates to the relative weight assigned to economic versus animal welfare considerations. For example, even though dairy farmers are generally aware that keeping cows in poorly bedded cubicles results in hock lesions, some choose to keep animals in such conditions for economic reasons. They may even claim that they have no choice because they would not survive economically if they were to use more costly bedding.

The chapters in this book provide no easy solutions to such dilemmas, but we hope that they will make readers think about the topics covered and allow them to recognize better some of the complexities involved. Each chapter ends with a few questions designed to facilitate discussion among readers. We believe that such discussion is valuable in itself, and it may also lead to ideas for how these and other dilemmas in animal welfare may eventually be resolved.

2

Tales about Tails: Is the Mutilation of Animals Justifiable in Their Best Interests or in Ours?

Sandra Edwards[1]* and Pauleen Bennett[2]
[1]University of Newcastle, UK; [2]La Trobe University, Australia

2.1 Abstract

Tail docking, involving surgical or non-surgical removal of a portion of the tail, is one of the most widely carried out and contentious mutilations inflicted by humans on animals. To differing extents, this procedure is carried out on farm livestock, draught animals and companion animals. The justifications range from benefits for the animals themselves, in reducing risk of future injury or disease, to human convenience or aesthetic preference. However, extensive scientific research indicates that the animals will experience some degree of acute pain and distress at the time of the procedure and medium-term pain arising from tissue damage, with longer-term chronic pain and adverse health effects also possible. As the acute pain can be controlled by the use of anaesthesia and analgesia and the absence of a tail has seldom been shown to disadvantage the animals greatly, a utilitarian analysis focusing on direct effects might conclude tail docking to be an acceptable procedure where significant benefits are obtained. However, it is important to consider whether, in condoning procedures that are justified as short-term solutions to existing suboptimal practices, we delay the implementation of more desirable longer-term solutions and potentially promote instrumental attitudes towards animals that we might prefer were discouraged.

*E-mail: sandra.edwards@ncl.ac.uk

2.2 Introduction

Mutilation of an animal is defined in English law as 'a procedure which involves interference with the sensitive tissues or bone structure of an animal, otherwise than for the purpose of its medical treatment' (Defra, 2007, p. 2). Such procedures have been widely carried out in the past on both farm livestock and companion animals. They include castration of male animals in all species and of female animals in many companion animal species, removal of horns and supernumerary teats of cattle, dew claws of dogs, and beak tips and claws of poultry. Removal of small segments of tissue from ears or feet, when tagging or notching for identification purposes, is also common in farm and laboratory animals. In recent times, all of these practices have become, to a greater or lesser extent, a subject of debate from both animal welfare and ethical perspectives.

One of the most widely carried out and contentious mutilations is tail docking (Bennett and Perini, 2003; Sutherland and Tucker, 2011). This involves amputation of part or all of the tail, by surgical severance or by necrosis following occlusion of the blood flow, in a procedure usually performed in young animals without anaesthesia or analgesia. Tail docking has, at some times and in some countries, been carried out routinely in farm livestock (cattle, sheep, pigs), draught animals (horses) and companion animals (dogs). A similar procedure is involved in tail resection for blood sampling purposes in laboratory rodents (BVA, 1993), but is not considered further in this chapter as it is a specific experimental procedure rather than routine commercial practice. The reasons for docking are different in different species and range from animal welfare concerns through to human convenience or even aesthetic preference. It has become perhaps one of the most extensively debated of the mutilations because of its widespread performance, the seemingly innocuous nature of the procedure itself, the importance attached to the benefits of its implementation in some species and, in the case of companion dogs, personal involvement by numerous members of the general public who rarely engage directly with other animals or concern themselves with welfare debates.

In this chapter, using tail docking as an example, we consider whether it is ever ethically acceptable to carry out mutilations and, if so, in which situations the benefits arising from such a procedure might be deemed to outweigh the harm involved.

2.3 Ethical Approaches to Consideration of Mutilations

The consideration of whether or not mutilations should be performed requires an ethical framework within which the available evidence can be placed. In a consideration of the ethics of neutering companion animals, which has many parallels with the issues of tail docking, Palmer *et al.* (2012) discussed how contrasting ethical approaches might give rise to different conclusions.

Rights approaches would suggest that any animal has a right not to be harmed and to be treated with respect. Thus, if mutilation imposes non-trivial harm, such as pain, distress or health compromise associated with the procedure, no future possibility of beneficial consequences will justify the violation of these rights. A particular consideration in relation to mutilations is the right to bodily integrity. While humans can relinquish such rights voluntarily if they perceive significant welfare benefits from doing so, animals do not have this capability and it is problematic for humans to make this choice on their behalf. The concept of integrity has been described by Yeates *et al.* (2011, p. 425) as 'resting upon a pre-scientific understanding that does not reduce the animal solely to an object for human use but also sees it as another being-in-flesh as ourselves'. In this regard, the effect (or lack of effect) on animal welfare is not pivotal to the central issue of the rightness of the act itself.

In contrast, consequentialist approaches, of which utilitarianism is one theoretical subset, consider that we should act to bring about the best outcomes in terms of maximizing good and minimizing that which is bad. This would imply that, while performing painful or distressing procedures on animals without benefit to themselves or others is clearly unacceptable, possible benefits from averting disease, pain or distress later in life might be sufficient to outweigh the lesser harm of the early mutilation. However, a range of other possible longer-term harms must also be considered, such as would arise if, as a result of the mutilation, animals were deprived of the future possibility of expressing important natural capacities or were more likely to be kept in situations which would impose other welfare harm. We must also consider whether, as a result of the mutilation, the human–animal relationship may be affected in a way that will impact, either positively or negatively, on the well-being of the parties concerned.

In utilitarian considerations, it is necessary to make decisions after balancing the, sometimes conflicting, interests of the different parties involved. This is more challenging than it might first appear, since it might be considered that human interests should predominate, that the interests of different parties should be weighted equally or that certain interests of animals can never be overridden for the sake of human or common good. In the rest of this chapter, we have assumed a consequentialist framework in that we focus only on outcomes in terms of animal welfare. However, we must leave it to the reader to decide how the outcomes should be weighed in the final moral equation.

2.4 The Benefits Accruing from Tail Docking

The justification for tail docking in different species generally falls into one of two categories. The most persuasive justification, for most people who give priority to animal welfare over economic or aesthetic human preferences, relies on outcomes which are beneficial for the animal itself. However, there may also be outcomes which give benefit to humans but do not in any way improve, or even detract from, the life of the animal.

2.4.1 Benefits to the animal

The most common justification for tail docking is to reduce the risk of future injury to the animal. In the case of pigs, this is to reduce the risk of being tail bitten by other group members. Tail biting is a widespread abnormal behaviour in intensively farmed pigs, affecting as many as 5% of all animals, though seen less frequently in animals housed more extensively (EFSA, 2007). It is a complex behaviour with many causal factors, including lack of enrichment, dietary inadequacy, deficiency in resource provision, climatic inadequacy, genetic predisposition and poor health (Schrøder-Petersen and Simonsen, 2001; Taylor *et al.*, 2010; Edwards, 2011). Despite the fact that most of these have been known for many years, the risk factors are still widespread on commercial farms (Taylor *et al.*, 2012). Animals which are tail bitten suffer serious consequences that escalate from mild discomfort, through minor injury to amputation of the whole tail and continuing cannibalism up into the spine (Schrøder-Petersen and Simonsen, 2001; Taylor *et al.*, 2010). Infections can enter the open wound readily and, even after intervention and healing, long-term pathologies such as pyaemia (blood poisoning characterized by pus-forming microorganisms in the blood) and lung abscesses frequently result.

The value of tail docking in reducing the risk of tail biting has sometimes been questioned (Moinard *et al.*, 2003), but countries which have implemented a ban on tail docking have an abattoir prevalence of bitten pigs which is 3–4 times higher than that in countries where most pigs are docked (EFSA, 2007), while comparisons of docked and undocked pigs at the same abattoir have shown a similar differential (Hunter *et al.*, 1999, 2001). In fact, this probably underestimates the likely benefit, since countries with a ban in place generally have less risky housing systems due to a requirement for straw provision, while farmers voluntarily choosing not to dock are likely to be those who perceive a low-risk situation. There are relatively few controlled studies which have examined tail injuries associated with docked and entire tails under the same conditions. However, three comparative studies in unbedded systems have shown an increase in the prevalence of bitten pigs from <10% in docked animals to >50% in those left with entire tails (Edwards, 2011). The most convincing comparison, recently replicated across four different farms in Denmark, showed a progressive increase in tail biting prevalence related to the length of tail remaining after docking (Thodberg *et al.*, 2010). Log odds ratios for tail biting in pigs docked leaving 5.7 cm, 7.5 cm and intact tails were 2.8, 3.3 and 4.6 compared to pigs short-docked to a 2.9 cm tail. While the evidence suggests that, under current farming conditions, tail docking of pigs provides a very significant protective benefit, such comparisons are not simple to interpret as pigs with less tail remaining have less tail to lose. Furthermore, permitted continuation of tail docking in pigs, to reduce risk of tail biting in current farm systems, may discourage moves towards better, but sometimes more expensive, systems in which the risks for undocked animals would be lower. It may also permit the continued selection of genetic lines with higher production merit but greater predisposition to perform the injurious behaviours, thus making any future attempt to cease docking still more difficult.

In the case of sheep, the primary justification for tail docking is to prevent fly strike, during which eggs are deposited on the animal and the hatched larvae burrow into the tissues. The prevalence of animals affected can vary from <1 to almost 20% (reviewed by Sutherland and Tucker, 2011), with affected individuals showing increased restlessness, decreased feeding, weight loss and, in cases of serious infestation, severe tissue damage and secondary infections. It is believed that tail docking reduces the risk of fly strike by preventing the accumulation of faecal material around the hindquarters, a demonstrated risk factor for fly strike (French *et al.*, 1994; Leathwick and Atkinson, 1995). However, the relationship between docking and faecal soiling is controversial; while some studies report greater soiling with long tails (Scobie *et al.*, 1999; Fisher and Gregory, 2007), others report no difference or more soiling with very short tails (Watts and Marchant, 1977; Scobie *et al.*, 1999). Furthermore, direct comparison of fly strike prevalence in docked and undocked animals on the same farm has given inconsistent results (French *et al.*, 1994; Webb Ware *et al.*, 2000). Thus, while strongly believed to be protective by many farmers, the evidence for the benefits of docking is not conclusive.

In dairy cattle, tail docking has been suggested to improve udder health by preventing the transfer of faecal material from a soiled tail to the udder. However, critical studies have shown that docked cows have neither cleaner udders (Tucker *et al.*, 2001; Schreiner and Ruegg, 2002b; Lombard *et al.*, 2010) nor lower prevalence of mastitis (Tucker *et al.*, 2001; Schreiner and Ruegg, 2002b). Evidence of a benefit of docking for the animal itself is therefore completely lacking in this species.

In draught horses, it has been suggested that tail docking may be beneficial to the animals themselves in reducing the risk of penis injury during mating or the risk of infection following gynaecological examination, although such suggestions have little scientific support (Lefebvre *et al.*, 2007).

In dogs, justification for tail docking takes many forms (Bennett and Perini, 2003) but is often based on reduced risk of tail injury (Orlans *et al.*, 1998). Many studies show tail injury to be rare (e.g. <0.5%; Darke *et al.*, 1985) and not significantly greater in undocked than docked animals. However, in the case of working dogs, a significant association between tail injuries and being undocked has been demonstrated in some gundog breeds (Houlton,

2008). Such results are difficult to interpret, as a dog with a tail is inevitably more likely to damage that tail than one without. Obviously, if the right hind leg of each animal in a group was amputated, they would be less likely to damage this limb subsequently than a control group with intact legs! It is of interest, therefore, that a more recent UK study (Diesel *et al.*, 2010), on a population of 138,000 dogs of different classes, showed a prevalence of tail damage of just 0.23%. While dogs with undocked tails were significantly more likely to sustain a tail injury, these authors calculated that approximately 500 dogs would need to be docked in order to prevent one tail injury. Thus, the benefit to the animals, while present, appears to be minor, except in a small number of situations, such as for spaniels working as gundogs. Additionally, however, some breeds of dogs carry a mutation (NBT) that results in a 'naturally bobbed' tail (Fig. 2.1). Individual dogs that inherit this NBT mutation are sometimes born with misshapen or deformed tails that may occasionally result in ongoing pain or be more prone to damage. Selective docking of these dogs soon after birth can prevent later problems in adult dogs. While the gene responsible could be removed through selective breeding, this would have the unwelcome consequence of markedly reducing already small gene pools.

Fig. 2.1. A litter of puppies in which some have the NBT gene, resulting in short tails.

2.4.2 Human benefits

While the benefits to animals from tail docking are sometimes questionable, benefits to their human owners are often clear. Docked pigs may be kept with lower risk of tail biting in more intensive housing systems that have a lower cost of production, though may be considered to provide less well for some other welfare requirements of the animals. The losses associated with tail biting in pigs as a result of mortality, veterinary treatment, lost performance, unfavourable sale weight and carcass condemnation can be so great – as much as 25% of the total sale value of the batch in which a tail biting outbreak occurs (Edwards, 2012) – that docking seems a justifiable precaution on both economic and animal welfare grounds. The emotional trauma to the stockperson of experiencing an outbreak of this severity and seeing the animals suffer should also not be underestimated. Similar economic losses and emotional trauma are associated with an outbreak of fly strike in sheep. Moreover, an additional practical human benefit of having docked sheep may be reduced shearing time (Scobie *et al.*, 1999), which may also benefit the animals indirectly.

In the case of dairy cattle, the major perceived benefit of tail docking is that of worker comfort. While the improved job satisfaction resulting from the reduced probability of being struck by a tail, sometimes manure coated, during milking is largely anecdotal, farmers that dock tails certainly perceive this to be important (Barnett *et al.*, 1999).

In draught horses, the most commonly cited justification for tail docking is one of handlers' safety, as the tail may interfere with the function of the reins used to guide the horse during agricultural or forestry work (Lefebvre *et al.*, 2007). However, these authors point out that undocked horses are widely used for draught activities in many countries without apparent problems, suggesting that such justification is not well founded and that the docking of certain breeds may now be more a matter of tradition and aesthetic reasons (Cregier, 1990).

It is in the case of docking in dogs where these less tangible human benefits play the greatest role. Tail docking in many breeds is an established custom and became part of the aesthetic requirement for these breeds when written breed standards were developed. In situations where docking is legal, the choice about whether or not to dock is left to individual breeders, many of whom remain deeply committed to preserving the established appearance of docked breeds (Orlans *et al.*, 1998). Moreover, since failure to dock may preclude pedigree animals from success in shows, thereby affecting their financial value and reproductive opportunities, normative pressure on breeders to continue with the established practice of docking can be intense. This is also an issue for breeders who transport dogs internationally. Dogs from traditionally docked breeds, produced by breeders in countries where docking has been banned, may be at a disadvantage when competing internationally.

A global ban on the docking of dogs for aesthetic reasons would ameliorate this competitive disadvantage. A second complication, however, is that, while many breeders have devoted a lifetime of effort to producing dogs that look 'just right', in breeds that carry the NBT gene, responsible for producing a 'naturally bobbed' tail, different breed lines may be affected differentially. While some breeders may be largely unaffected by a ban on tail docking, others may elect to remove many individual dogs from their breeding programme, including some that were previously highly valued. While this may appear to be a somewhat trivial consideration, it could have the effect of destroying the established aesthetic of the breed, putting some breeders 'out of business' and reducing already depleted canine gene pools. The presence of the NBT gene in a population also means that enforcement of a ban on docking is challenging. Following the introduction of legislation banning the docking of domestic dogs in Australia, litter registrations in some previously docked breeds were substantially reduced. In other breeds, it remains impossible for compliance officers to determine if individual dogs have been surgically docked or carry a 'naturally bobbed' tail. The NBT mutation often results in puppies being born with shortened or misshapen tails. As this is sufficient for some veterinarians to justify docking on prophylactic grounds, some breeders have elected to introduce the NBT mutation into their lines, rather than eliminating it in favour of full-length tails, the docking of which is illegal.

2.5 The Harm Associated with Tail Docking

The direct harm associated with tail docking can be considered under various categories. The first is the acute pain and stress associated with the amputation procedure itself; the second is the medium-term pain associated with inflammation or necrosis, depending on the amputation method; the third is the possibility of long-term chronic pain as a result of nerve damage and neuroma formation; and the final issue is whether the absence of the tail has adverse effects on the subsequent life of the animal concerned. In addition to direct harm, tail docking may be associated with indirect harm for animals in that its acceptance by a community in the absence of a sound empirical justification may desensitize that community to similar, or more damaging, procedures performed on animals.

2.5.1 Acute pain

In all species, the amputated section of tail comprises skin, muscle, bone and nerves. It is therefore to be expected that any damage would be a source of pain to the animal, although the evidence of this is not always strong. In the case of the pig, where amputation is performed in the first week of life by surgical resection, carried out with pliers, scissors, a scalpel blade or a cautery iron, behavioural signs of pain and distress during the procedure are manifested by increased struggling and vocalizations which are greater in number and frequency (pitch) than those of control animals subject to a sham operation (Noonan *et al.*, 1994; Marchant-Forde *et al.*, 2009). However, this research finding is disputed by some farmers (personal communications to S.A. Edwards), who claim they can dock tails without any apparent signs of distress while piglets are suckling. The evidence from measurements of stress physiology is also conflicting, with some studies reporting no changes in adreno-corticotrophic hormone (ACTH), cortisol, lactate or β-endorphin levels in response to docking (Prunier *et al.*, 2005; Sutherland *et al.*, 2008; Marchant-Forde *et al.*, 2009), while other studies report cortisol levels elevated above those of control-handled piglets for some docking methods only (Sutherland *et al.*, 2008, 2011). This clearly cannot be explained only by the differences between amputation methods, however, as comparison of plier and cautery docking has resulted in different conclusions from different studies (Sutherland *et al.*, 2008; Marchant-Forde *et al.*, 2009).

In sheep and cattle, a wider variety of amputation methods is used in tail docking (Sutherland and Tucker, 2011). While the surgical methods used in pigs are sometimes adopted, a more common method is the use of a constrictive rubber ring which occludes blood flow to the distal portion of the tail. This tissue then necrotizes over a period of weeks, before falling off or being removed. When surgical methods are used, either with or without cautery, both behavioural signs of acute pain (Kent *et al.*, 1998; Grant, 2004) and elevated blood cortisol concentrations (Lester *et al.*, 1991; Morris *et al.*, 1994) are seen in lambs. The same behavioural and physiological evidence of acute pain is shown in response to rubber ring application (Morris *et al.*, 1994; Kent *et al.*, 1998). While the docking method adopted can result in a different pattern and time course of behavioural and physiological responses (Morris *et al.*, 1994; Lester *et al.*, 1996), there is little doubt that all induce significant acute pain. In the case of cattle, the evidence is less clear. Behavioural changes after application of the rubber ring are either absent (Eicher *et al.*, 2000; Eicher and Dailey, 2002) or less extreme in calves (Petrie *et al.*, 1995; Tom *et al.*, 2002a) or older cattle (Tom *et al.*, 2002b) than in lambs. The same uncertainty is seen in physiological measures, with published studies only sometimes showing changes in cortisol concentration (Eicher *et al.*, 2000; Tom *et al.*, 2002a) or showing no adrenal response in heart and respiration rate (Schreiner and Ruegg, 2002a). In horses, both surgical and non-surgical methods of docking are also used (Lefebvre *et al.*, 2007), but no scientific study of the associated pain appears to have been carried out.

In dogs, tail docking is most often performed surgically, although rubber ring constriction is also used by some owners (Noonan *et al.*, 1996a). It generally takes place within the first few days of life, and without anaesthesia or analgesia. Although many breeders consider the practice to cause little or no pain (Noonan *et al.*, 1996a), the limited evidence available indicates that this is not the case. Docked puppies struggle and vocalize intensely at the time of surgical amputation and immediately afterwards (Noonan *et al.*, 1996b), although these behavioural changes appear very transient. While there have been no reported studies of how puppies react acutely to rubber ring constriction, an added complication with respect to this procedure is that it has typically been performed by dog breeders, with varying levels of experience, rather than by qualified veterinarians. While inexperienced breeders can locate instructions for docking easily on the Internet, lack of training, supervision and experience may result in poor placement of the rubber ring, potentially leading to increased pain or stress for the animal.

It has been suggested that the lack of a physiological stress response to tail docking in young animals might simply reflect the fact that the pituitary–adrenocortical axis is relatively immature and not responsive to stress at this time (Prunier *et al.*, 2005). However, data from ACTH challenge in piglets as young as 3 days old show this not to be the case (Otten *et al.*, 2001). It has also been suggested that newborn animals may not have fully developed conscious perceptions of pain in the immediate post-natal period and that procedures carried out on the very young animal are therefore more acceptable. For example, it has been demonstrated that very young lambs have a reduced cerebrocortical response to castration (Johnson *et al.*, 2009). This may be even more likely in dogs, in which, as an altricial species, the young are born in a much less developed state, but no canine data are available with which to address this issue. Although it has been shown that endogenous neuroinhibitory mechanisms restricting consciousness, which function in the fetal state, persist for a period after birth (Mellor and Diesch, 2006), the behavioural evidence indicates that these mechanisms are no longer protective in the species considered here at the time when docking is normally carried out. More recent research also highlights how the perception of pain from these procedures might be influenced by the previous experience of the individual concerned. Thus, prenatal stress reduced sensitivity to noxious stimulation applied at the base of the tail in piglets (Sandercock *et al.*, 2011), while prior castration increased the behavioural response of lambs to tail docking (McCracken *et al.*, 2010). Emotional and physiological states have been shown to alter pain perception in many studies (Watkins and Maier, 2002), but neither prior isolation stress nor immune activation was shown to influence pain thresholds or endocrine responses to castration and tail docking in lambs (Clark *et al.*, 2011).

The acute pain associated with tail docking might be abolished by the use of anaesthesia or analgesia at this time. Although one study reported that topical analgesia can reduce the behavioural signs of pain in piglets (Prunier *et al.*, 2001), another study found neither concurrent topical, injected nor general anaesthesia using CO_2 significantly changed the acute physiological or behavioural response to tail docking. However, in this study, a short-acting topical anaesthetic did give some attenuation of the cortisol response and of changed lying patterns in the 2 h post docking, suggesting the possibility of some medium-term pain alleviation (Sutherland *et al.*, 2011). Topical anaesthesia has also been reported to reduce behavioural signs of pain after hot iron docking in lambs (Lomax *et al.*, 2010), although this approach is less effective than local anaesthesia and analgesia to alleviate the pain of ring docking (Graham *et al.*, 1997; Kent *et al.*, 1998). In dairy calves, the use of local anaesthesia before tail banding eliminated behavioural reactions which might be indicative of pain (Petrie *et al.*, 1995) but, in studies of older animals, the absence of symptoms in either the presence or absence of epidural anaesthesia made conclusions on the value of pain relief difficult (Schreiner and Ruegg, 2002a).

2.5.2 Medium-term pain

Pain during the hours and days following docking would be expected to occur as a result of tissue damage giving rise to inflammation following surgical amputation, or necrosis following ring amputation. Signs of pain in the period after docking has taken place can be observed as tail wagging, tail jamming or scooting for up to 45 min in piglets (Sutherland *et al.*, 2008). No behavioural studies have looked in detail at the subsequent hours or days, although one study reported reduced growth rates to 14 days of age in piglets docked by cautery but not those docked by pliers (Marchant-Forde *et al.*, 2009). Haematology and C-reactive protein, an acute-phase protein produced in response to infection, inflammation or trauma, did not differ in docked and undocked piglets at 3 weeks of age (Sutherland *et al.*, 2009), indicating no residual inflammatory response by this time.

In the case of lambs, active pain behaviours and abnormal postures in the period immediately after cautery docking show little difference from controls (Lester *et al.*, 1991, 1996; Graham *et al.*, 1997; Grant, 2004), although a small increase in active pain behaviours has been observed between 30 and 60 min, which is consistent with the delayed onset of burn pain reported in humans. It has been suggested that a reduced pain response following cautery may occur as a result of the destruction of nociceptors at the tail stump (Lester *et al.*, 1991). In contrast, ring docking results in large behavioural and postural responses lasting for at least 90 min (Grant, 2004), suggested to result from the occlusion of blood supply but not prevention of nerve impulses from the ischaemic tail tissue (Molony *et al.*, 1993; Kent *et al.*, 1995). Destruction of the nerves by crushing before or after the ring is applied can alleviate this pain, as indicated by less time spent in pain behaviours and reduced cortisol concentration at 1 h after docking (Graham *et al.*, 1997; Kent *et al.*, 1998), but the crushing procedure itself causes further acute pain.

In cattle, detailed studies for the first 12 h after ring docking, and continuing studies up to 6 weeks, detected no medium-term changes in posture or eating, drinking and ruminating behaviour (Schreiner and Ruegg, 2002a).

There were also no significant changes in haematological data up to 21 days related to treatment.

It appears that puppies also recover quickly from the acute trauma of surgical docking or the application of a rubber ring and soon suckle or fall asleep. However, interpretation of the quiescence which follows docking is difficult, as young puppies have an extremely limited behavioural repertoire. While cessation of acute pain behaviours may be indicative of no further pain, it could also reflect an adaptive mechanism to deal with pain or exhaustion (Bennett and Perini, 2003). Sleep often follows a stressful experience in young infants, while suckling behaviour may reduce stress (hence the widespread use of pacifiers to sooth human babies) or provide analgesia by stimulating the release of endogenous opioids. Without more detailed studies, this question currently cannot be resolved.

In studies of medium-term pain, few detailed investigations have continued beyond 2–3 h post docking, although Herskin *et al.* (2012) reported changes in the lying behaviour of piglets which were related systematically to tail docking length and continued until at least 6 h after docking, leaving open the possibility of continuing discomfort from tissue trauma and inflammatory responses.

2.5.3 Long-term chronic pain

By far the least understood aspect of tail docking is the possibility of long-term chronic pain. This has been suggested by the observation of neuromas in the healed tail stumps of pigs (Simonsen *et al.*, 1991), lambs (French and Morgan, 1992), cattle (Eicher *et al.*, 2006) and a small sample of dogs (Gross and Carr, 1990). Neuromas are masses of neural tissue which form when axons are severed. In humans, they can be associated with hypersensitivity and spontaneous pain (Lewin-Kowalik *et al.*, 2006), but the expression of this varies greatly between individuals. In animals, there is uncertainty about their significance. Despite attempts, there are few studies that have been able to demonstrate increased sensitivity in healed tail stumps, although in one study tail-docked heifers showed increased sensitivity to heat and cold (Eicher *et al.*, 2006). Anecdotal reports of sensitive tail stumps in dogs have also been published (Bennett and Perini, 2003).

A recent study in piglets investigated molecular markers in the coccygeal dorsal root ganglia and spinal cord indicative of nerve injury and regeneration, and of chronic pain, at 7 and 52 days post docking. Increase in these markers was seen at 7 days after tail docking but was not apparent by 52 days, suggesting that the effects on peripheral and spinal nociception were relatively short lasting (Sandercock *et al.*, 2012).

A further area of concern is the possibility that early nerve damage might cause changes in the developing nervous system, which would give rise to a general increase in the perception of pain in later life, as demonstrated in human infants and in sheep (Mellor *et al.*, 2010). Such a possibility has, to date, been little studied in relation to tail docking, although there is one study demonstrating that neonatal tail docking does not lead to long-term alterations in nociception in pigs (Sandercock *et al.*, 2011).

There may be other medium- and long-term health consequences arising from tail docking, although data are sparse. As tails are seldom docked under aseptic conditions, the risk of infection through any open wound is significant, although the incidence of negative post-operative side effects in commercial practice has not been assessed systematically in any species. Suggested adverse long-term health effects of tail docking in dogs include atrophy and degeneration of the tail and pelvic muscles, leading to faecal incontinence or perineal hernia (Wansborough, 1996). An association between tail docking and acquired urinary incontinence has also been reported (Holt and Thrusfield, 1993). However, there is a lack of scientifically controlled studies to evaluate these issues critically.

2.5.4 Absence of a tail

As many different species have evolved with a tail, it can be assumed that this appendage must serve some useful function in an evolutionary context. The extent to which the lives of docked animals are therefore impaired by the absence of a complete tail has perhaps received less study than it should have. Known roles for the tail include balance during motion, the removal of insects from the body and, perhaps most importantly, a major role in intra- and inter-species communication (Hickman, 1979).

The role of the tail in balance would appear to be relatively minor in most farm species, where the mass of the body is relatively great in comparison to the tail and where opportunities for the types of movement engaged in by ancestral stock are curtailed. However, in the case of the dog, kinematic studies have shown that tail movements are important in maintaining body balance during locomotion (Wada *et al.*, 1993). Despite this, there are many examples of success of docked dogs in agility competitions and other demanding sporting and working activities (Bennett and Perini, 2003), suggesting that any detrimental effects of tail absence are relatively minor.

The role of the tail in insect removal has been studied in cattle, though similar studies in pigs, sheep, horses or dogs are lacking. Docked cattle have more flies on the hind regions of their body than undocked animals (Phipps *et al.*, 1995; Eicher *et al.*, 2001; Eicher and Dailey, 2002). Inability to remove biting flies will result in increased discomfort and has the potential to increase susceptibility to insect-borne diseases.

The nature of tail movements and their display function was described for many species, including pigs, cattle, horses and dogs, by Kiley-Worthington

(1976). While the significance of the tail for communication has not been studied experimentally in farm livestock, it has been widely demonstrated that the domestic dog can express social status and emotional state by changes in the position and motion of the tail. The consequences of tail docking for intra-species communication in the dog have been studied experimentally, using a model animal with a differing length of tail, by Leaver and Reimchen (2008). They demonstrated that smaller dogs approached the docked model more cautiously, while larger dogs stopped more frequently during their approach, suggesting uncertainty about the social cues. Furthermore, tail-wagging movements were responded to appropriately for a long tail but not differentiated when the tail was docked, showing impaired signal communication. This suggests the potential for increased aggression due to social misunderstanding when tails are docked, an outcome reported anecdotally by Coren (2000). The docking of tails in dogs also has potential consequences for interspecies communication. The inability of humans to understand correctly the warning signals given by a dog is a major cause of injury from canine aggression (Mertens, 2002). It has been shown in one study that tail movements are the most common cues used by humans with different degrees of dog experience to interpret dog behaviour (Tami and Gallagher, 2009) and that aggression is poorly classified in comparison to other behavioural states, but another study has found that owners reported relying on vocalizations and gross body movements more than tail activity when attempting to interpret their dog's emotional state (Kerswell *et al.*, 2009).

2.5.5 Indirect harm

It has long been argued that significant indirect harm associated with any form of violence towards animals is the potential for engagement in such activities to lead to further mistreatment, perhaps of a more serious nature, of humans or other animals (Flynn, 2011). In the contemporary literature, this hypothetical tendency is most often discussed in relation to what is now widely referred to as the 'link' between the perpetration of animal abuse and various forms of both violent and non-violent criminal activity (Ascione, 2008). While the 'link' receives much attention in the popular press, the presence of a causal relationship between animal abuse and criminal activity has been contested in the scientific literature (Patterson-Kane and Piper, 2009). In addition, while there is some evidence to suggest that slaughterhouse employment may be associated with various social problems (Fitzgerald *et al.*, 2009), the extent to which engagement in socially-sanctioned mutilations, such as tail docking, may lead to a broader expression of violence by individuals has not been investigated adequately. Nor is it clear whether, in condoning any form of mutilation of animals, we may potentially promote societal attitudes towards animals that are inappropriate in the modern era.

2.6 Assessing the Cost–Benefit Balance

From the foregoing discussion, it can be concluded that tail docking causes some degree of acute and medium-term pain, though to a differing extent in different species and depending on which method of docking is employed. Unfortunately, it remains impossible to quantify this pain precisely, as the subjectivity of pain renders it largely invisible to scientific instruments. While methods for inferring pain in non-verbal animals have advanced in recent years, the applicability of these methods to very young animals, particularly those with limited behavioural responses and undeveloped physiological responses, is uncertain.

To some extent, the issue of acute and medium-term pain is ameliorated by the fact that pain can be reduced by the appropriate use of anaesthesia and analgesia, although this is currently rare and optimal methods have yet to be developed. There is little evidence for any long-term pain consequent on docking, and the subsequent absence of a tail, though giving rise to some behavioural deficit, does not seem to disadvantage most animals greatly. This would suggest that a utilitarian analysis focusing on the direct effects would conclude tail docking to be an acceptable procedure if significant benefits are obtained. An exception to this may be domestic dogs, whose tails may facilitate the interspecies communication with humans and so be more important for their survival and well-being.

A consequentialist approach would suggest that the benefits which should be given greatest consideration are those relating to the welfare of the animal itself and not to human financial profit or aesthetic whim. These vary in different species and, while they appear to justify tail docking in some circumstances, in considering the cost–benefit balance it is important not to enshrine practices which may be in the best interests of animals in the short term but which are not in their best long-term interests, as may be the case with tail docking in pigs.

This, then, brings us back to the issue of whether, in condoning procedures that are justified only because of existing sociocultural practices, we potentially promote instrumental attitudes towards animals that we might prefer were discouraged. Clearly, there is a strong prima facie case against the mutilation of animals. Just as clearly, removal of an animal's tail constitutes mutilation as defined under law in England and other countries, even if any resulting pain is minimal and well controlled, which is rarely the case. Whether exceptions to this law should continue to be sanctioned, therefore, requires ongoing consideration.

2.7 Conclusions

- Tail docking is a mutilation which is widely carried out in many species and appears relatively innocuous. The justifications range from benefits for the animals themselves, in reducing risk of future injury or disease, to human convenience or aesthetic preference.

- There is clear evidence of acute pain and distress in most species at the time of the procedure and medium-term pain arising from tissue damage, with longer-term chronic pain and adverse health effects also possible.
- As the acute pain can be controlled by the use of anaesthesia and analgesia, and the absence of a tail has seldom been shown to disadvantage the animals greatly, a utilitarian analysis focusing on direct effects might conclude tail docking to be an acceptable procedure where demonstrable and significant benefits are obtained.
- Condoning tail docking as a short-term solution to prevent harm arising from existing suboptimal practices may delay the implementation of more desirable longer-term solutions and potentially promote instrumental attitudes towards animals that we might prefer were discouraged.

Questions for discussion

1. You are the veterinarian for a farm where tail docking of pigs is currently carried out. Should you advise your client to cease this practice immediately?
2. Should you purchase a puppy that you particularly want, even if it has been tail docked? Assume that this procedure is legal in your country.
3. Should we remove the need for tail docking by focused genetic selection against expression of the problem which prompts this procedure?
4. In some countries, for example Sweden, there is a legal ban on tail docking of both dogs and pigs, except in cases where the animal is injured. Is this the way forward in other countries too?

Acknowledgements

We are grateful for contributions to the chapter by Peter Sandøe and the other authors of this book.

References

Ascione, F.R. (ed.) (2008) *The International Handbook of Animal Abuse and Cruelty: Theory, Research and Application*. Purdue University Press, West Lafayette, Indiana.

Barnett, J.L., Coleman, G.J., Hemsworth, P.H., Newman, E.A., Fewings-Hall, S. and Ziini, C. (1999) Tail docking and beliefs about the practice in the Victorian dairy industry. *Australian Veterinary Journal* 77, 742–747.

Bennett, P.C. and Perini, E. (2003) Tail docking in dogs: a review of the issues. *Australian Veterinary Journal* 81, 208–218.

BVA (1993) Removal of blood from laboratory mammals and birds. First report of the BVA/FRAME/RSPCA/UFAW Joint Working Group on Refinement. *Laboratory Animals* 27, 1–22.

Clark, C., Mendl, M., Jamieson, J., Arnone, A., Waterman-Pearson, A. and Murrell, J. (2011) Do psychological and physiological stressors alter the acute pain response to castration and tail docking in lambs? *Veterinary Anaesthesia and Analgesia* 38, 134–145.

Coren, S. (2000) *How To Speak Dog*. Free Press, New York.

Cregier, S.E. (1990) Shocking docking: mutilation before education? *Equine Veterinary Science* 10, 252–255.

Darke, P.G., Thrusfield, M.V. and Aitken, C.G. (1985) Association between tail injuries and 533 docking in dogs. *Veterinary Record* 116, 409.

Defra (2007) The Mutilations (Permitted Procedures) (England) Regulations 2007. SI 2007 No 1100. Department for Environment, Food and Rural Affairs, London.

Diesel, G., Pfeiffer, D., Crispin, S. and Brodbelt, D. (2010) Risk factors for tail injuries for dogs in Great Britain. Report to Defra.

Edwards, S.A. (2011) What do we know about tail biting today? *Pig Journal* 66, 81–86.

Edwards, S.A. (2012) Tail biting in pigs – an international overview. In: Proceedings 17 Internationale Fachtagung zum Thema Tierschutz, Nurtingen, 12–13 March 2012. Deutsche Veterinarmedizinische Gesellschaft eV, Giessen, Germany, pp. 11–26.

EFSA (European Food Safety Authority) (2007) Scientific report on the risks associated with tail biting in pigs and possible means to reduce the need for tail docking considering the different housing and husbandry systems. *EFSA Journal* 611, 1–98.

Eicher, S.D. and Dailey, J.W. (2002) Indicators of acute pain and fly avoidance behaviors in Holstein calves following tail-docking. *Journal of Dairy Science* 85, 2850–2858.

Eicher, S.D., Morrow-Tesch, J.L., Albright, J.L., Dailey, J.W., Young, C.R. and Stanker, L.H. (2000) Tail-docking influences on behavioral, immunological, and endocrine responses in dairy heifers. *Journal of Dairy Science* 83, 1456–1462.

Eicher, S.D., Morrow-Tesch, J.L., Albright, J.L. and Williams, R.E. (2001) Taildocking alters fly numbers, fly-avoidance behaviors, and cleanliness, but not physiological measures. *Journal of Dairy Science* 84, 1822–1828.

Eicher, S.D., Cheng, H.W., Sorrells, A.D. and Schutz, M.M. (2006) Behavioral and physiological indicators of sensitivity or chronic pain following tail docking. *Journal of Dairy Science* 89, 3047–3051.

Fisher, M.W. and Gregory, N.G. (2007) Reconciling the differences between the length at which lambs' tails are commonly docked and animal welfare recommendations. *Proceedings of the New Zealand Society for Animal Production* 67, 32–38.

Fitzgerald, A.J., Kalof, L. and Dietz, T. (2009) Slaughterhouses and increased crime rates: an empirical analysis of the spillover from 'the jungle' into the surrounding community. *Organization and Environment* 22, 158–184.

Flynn, C.P. (2011) Examining the links between animal abuse and human violence. *Crime Law and Social Change* 55, 453–468.

French, N., Wall, R., Cripps, P.J. and Morgan, K.L. (1994) Blow fly strike in England and Wales: the relationship between prevalence and farm and management factors. *Medical and Veterinary Entomology* 8, 51–56.

French, N.P. and Morgan, K.L. (1992) Neuromas in docked lambs' tails. *Research in Veterinary Science* 52, 389–390.

Graham, M.J., Kent, J.E. and Molony, V. (1997) Effects of four analgesic treatments on the behavioural and cortisol responses of 3-week-old lambs to tail docking. *Veterinary Journal* 153, 87–97.

Grant, C. (2004) Behavioural responses of lambs to common painful husbandry procedures. *Applied Animal Behaviour Science* 87, 255–273.

Gross, T.L. and Carr, S.H. (1990) Amputation neuroma of docked tails in dogs. *Veterinary Pathology* 27, 61–62.

Herskin, M.S., di Giminiani, P., Damgaard, B.M. and Thodberg, K. (2012) Tail docking in piglets – does increased docking length lead to increased behavioural responses? In: *Abstracts of the 14th World Congress on Pain, Milan, August 27–31, 2012*.International Association for the Study of Pain, Washington, DC, PT221.

Hickman, G.C. (1979) Mammalian tail – review of functions. *Mammal Review* 9, 143–157.

Holt, P.E. and Thrusfield, M.V. (1993) Association in bitches between breed, size, neutering and docking, and acquired urinary incontinence due to incompetence of the urethral sphincter mechanism. *Veterinary Record* 133, 177–180.

Houlton, J.E. (2008) A survey of gundog lameness and injuries in Great Britain in the shooting 540 seasons 2005/2006 and 2006/2007. *Veterinary Comparative Orthopedics and Traumatology* 21, 231–237.

Hunter, E.J., Jones, T.A., Guise, H.J., Penny, R.H.C. and Hoste, S. (1999) Tail biting in pigs 1: The prevalence at six UK abattoirs and the relationship of tail biting with docking, sex and other carcass damage. *Pig Journal* 43, 18–32.

Hunter, E.J., Jones, T.A., Guise, H.J., Penny, R.H.C. and Hoste, S. (2001) The relationship between tail biting in pigs, docking procedure and other management procedures. *Veterinary Journal* 161, 72–79.

Johnson, C.B., Sylvester, S.P. and Stafford, K.J. (2009) Effects of age on the electroencephalographic response to castration in lambs anaesthetised using halothane in oxygen from birth to six weeks old. *Veterinary Anaesthesia and Analgesia* 36, 273–279.

Kent, J.E., Molony, V. and Robertson, I.S. (1995) Comparison of the Burdizzo and rubber ring methods for castrating and tail docking lambs. *Veterinary Record* 136, 192–196.

Kent, J.E., Molony, V. and Graham, M.J. (1998) Comparison of methods for the reduction of acute pain produced by rubber ring castration or tail docking of week-old lambs. *Veterinary Journal* 155, 39–51.

Kerswell, K.J., Bennett, P., Butler, K.L. and Hemsworth, P.H. (2009) Self-reported comprehension ratings of dog behavior by puppy owners. *Anthrozoös* 22, 183–193.

Kiley-Worthington, M. (1976) Tail movements of ungulates, canids and felids with particular reference to their causation and function as displays. *Behaviour* 56, 69–115.

Leathwick, D.M. and Atkinson, D.S. (1995) Dagginess and fly strike in lambs grazed on *Lotus corniculatus* or ryegrass. *Proceedings of the New Zealand Society for Animal Production* 55, 196–198.

Leaver, S.D.A. and Reimchen, T.E. (2008) Behavioural responses of *Canis familiaris* to different tail lengths of a remotely-controlled life-size dog replica. *Behaviour* 145, 377–390.

Lefebvre, D., Lips, D., Odberg, F.O. and Giffroy, J.M. (2007) Tail docking in horses: a review of the issues. *Animal* 1, 1167–1178.

Lester, S.J., Mellor, D.J., Ward, R.N. and Holmes, R.J. (1991) Cortisol responses of young lambs to castration and tailing using different methods. *New Zealand Veterinary Journal* 39, 134–138.

Lester, S.J., Mellor, D.J., Holmes, R.J., Ward, R.N. and Stafford, K.J. (1996) Behavioural and cortisol responses of lambs to castration and tailing using different methods. *New Zealand Veterinary Journal* 44, 45–54.

Lewin-Kowalik, J., Marcol, W., Kotulska, K., Mandera, M. and Klimczak, A. (2006) Prevention and management of painful neuroma. *Neurologia Medico-Chirurfica (Tokyo)* 46, 62–67.

Lomax, S., Dickson, H., Sheila, M. and Windsora, P.A. (2010) Topical anaesthesia alleviates short-term pain of castration and tail docking in lambs. *Australian Veterinary Journal* 88, 67–74.

Lombard, J.E., Tucker, C.B., von Keyserlingk, M.A.G., Kopral, C.A. and Weary, D.M. (2010) Associations between cow hygiene, hock injuries, and free stall usage on US dairy farms. *Journal of Dairy Science* 93, 4668–4676.

Marchant-Forde, J.N., Lay, D.C., McCunn, K.A., Cheng, H.W., Pajor, E.A. and Marcghant-Forde, R.M. (2009) Postnatal piglet husbandry practices and well being: the effect of alternate techniques delivered separately. *Journal of Animal Science* 87, 1479–1492.

McCracken, L., Waran, N., Mitchinson, S. and Johnson, C.B. (2010) Effect of age at castration on behavioural response to subsequent tail docking in lambs. *Veterinary Anaesthesia and Analgesia* 37, 375–381.

Mellor, D.J. and Diesch, T.J. (2006) Onset of sentience: the potential for suffering in fetal and newborn farm animals. *Applied Animal Behaviour Science* 100, 48–57.

Mellor, D.J., Diesch, T.J. and Johnson, C.B. (2010) Should mammalian fetuses be excluded from regulations protecting animals during experiments? *Altex* 27, 199–202.

Mertens, P.A. (2002) Canine aggression. In: Horwitz, D., Mills, D. and Heath, S. (eds) *BSAVA Manual of Canine and Feline Behavioural Medicine*. British Small Animal Veterinary Association, Quedgeley, Gloucester, UK, pp. 195–215.

Moinard, C., Mendl, M., Nicol, C.J. and Green, L.E. (2003) A case control study of on-farm risk factors for tail biting in pigs. *Applied Animal Behaviour Science* 81, 333–355.

Molony, V., Kent, J.E. and Robertson, I.S. (1993) Behavioural responses of lambs of three ages in the first three hours after three methods of castration and tail docking. *Research in Veterinary Science* 55, 236–245.

Morris, D.G., Kuchel, T.R. and Maddocks, S. (1994) Stress responses in lambs to different tail docking methods. *Proceedings of the Australian Society for Animal Production* 20, 202–205.

Noonan, G.J., Rand, J.S., Priest, J., Ainscow, J. and Blackshaw, J.K. (1994) Behavioural observations of piglets undergoing tail docking, teeth clipping and ear notching. *Applied Animal Behaviour Science* 39, 203–213.

Noonan, G.J., Rand, J.S. and Blackshaw, J.K. (1996a) Tail docking in dogs: a sample of attitudes of veterinarians and dog breeders in Queensland. *Australian Veterinary Journal* 73, 86–88.

Noonan, G.J., Rand, J.S., Blackshaw, J.K. and Priest, J. (1996b) Behavioural observations of puppies undergoing tail docking. *Applied Animal Behaviour Science* 49, 335–342.

Orlans, F.B., Beauchamp, T.L., Dresser, R., Morton, D.B. and Gluck, J.P. (1998) Should the tail wag the dog? In: Orlans, F.B., Beauchamp, T.L., Dresser, R., Morton, D.B. and Gluck, J.P. (eds) *The Human Use of Animals. Case Studies in Ethical Choice*. Oxford University Press, New York, pp. 273–287.

Otten, W., Kanitz, E., Tuchscherer, M. and Nurnberg, G. (2001) Effects of prenatal restraint stress on hypothalamic-pituitary-adrenocortical and sympatho-adreno-medullary axis in neonatal pigs. *Animal Science* 73, 279–287.

Palmer, C., Corr, S. and Sandoe, P. (2012) Inconvenient desires: should we routinely neuter companion animals? *Anthrozoos* 25, S153–S172.

Patterson-Kane, E.G. and Piper, H. (2009) Animal abuse as a sentinel for human violence: a critique. *Journal of Social Issues* 65, 589–614.

Petrie, N.J., Mellor, D.J., Stafford, K.J., Bruce, R.A. and Ward, R.N. (1995) The behaviour of calves tail docked with a rubber ring used with or without local anaesthesia. *Proceedings of the New Zealand Society for Animal Production* 55, 58–60.

Phipps, A.M., Matthews, L.R. and Verkerk, G.A. (1995) Tail docked dairy cattle: fly induced behaviour and adrenal responsiveness to ACTH. *Proceedings of the New Zealand Society for Animal Production* 55, 61–63.

Prunier, P., Bataille, G., Meunier-Salaun, M.C., Bregeon, A. and Rugraff, Y. (2001) Conséquences comportementales, zootechniques et physiologiques de la caudectomie réalisée avec ou sans 'insensibilisation' locale chez le porcelet. *Journées de Recherche Porcine en France* 33, 313–318.

Prunier, A., Mounier, A.M. and Hay, M. (2005) Effect of castration, tooth resection, or tail docking on plasma metabolites and stress hormones in young pigs. *Journal of Animal Science* 83, 216–222.

Sandercock, D.A., Gibson, I.F., Rutherford, K.M.D., Donald, R.D., Lawrence, A.B., Brash, H.M., *et al.* (2011) The impact of prenatal stress on basal nociception and evoked responses to tail-docking and inflammatory challenge in juvenile pigs. *Physiology and Behavior* 104, 728–737.

Sandercock, D.A., Monteiro, A., Scott, E.M. and Nolan, A.M. (2012) The effect of tail-docking neonatal piglets on ATF-3 and NR2B immunoreactivity in coccygeal DRG and spinal dorsal horn neurons: preliminary data. *Scandinavian Journal of Pain* 3, 184–185.

Schreiner, D.A. and Ruegg, P.L. (2002a) Responses to tail docking in calves and heifers. *Journal of Dairy Science* 85, 3287–3296.

Schreiner, D.A. and Ruegg, P.L. (2002b) Effects of tail docking on milk quality and cow cleanliness. *Journal of Dairy Science* 85, 2503–2511.

Schrøder-Petersen, D.L. and Simonsen, H.B. (2001) Tail biting in pigs. *Veterinary Journal* 162, 196–210.

Scobie, D.R., Bray, A.R. and O'Connell, D. (1999) A breeding goal to improve the welfare of sheep. *Animal Welfare* 8, 391–406.

Simonsen, H.B., Klinken, L. and Bindseil, E. (1991) Histopathology of intact and docked pigtails. *British Veterinary Journal* 147, 407–412.

Sutherland, M.A. and Tucker, C.B. (2011) The long and the short of it: a review of tail docking in farm animals. *Applied Animal Behaviour Science* 135, 179–191.

Sutherland, M.A., Bryer, P.J., Krebs, N. and McGlone, J.J. (2008) Tail docking in pigs: acute physiological and behavioural responses. *Animal* 2, 292–297.

Sutherland, M.A., Bryer, P.J., Krebs, N. and McGlone, J.J. (2009) The effect of method of tail docking on tail-biting behaviour and welfare of pigs. *Animal Welfare* 18, 561–570.

Sutherland, M.A., Davis, B.L. and McGlone, J.J. (2011) The effect of local or general anesthesia on the physiology and behavior of tail docked pigs. *Animal* 5, 1237–1246.

Tami, G. and Gallagher, A. (2009) Description of the behaviour of domestic dog (*Canis familiaris*) by experienced and inexperienced people. *Applied Animal Behaviour Science* 120, 159–169.

Taylor, N.R., Main, D.C.J., Mendl, M. and Edwards, S.A. (2010) Tail-biting: a new perspective. *Veterinary Journal* 186, 137–147.

Taylor, N.R., Parker, R.M.A., Mendl, M., Edwards, S.A. and Main, D.C.J. (2012) Prevalence of tail biting risk factors on commercial farms in relation to intervention strategies. *Veterinary Journal* 194, 77–83.

Thodberg, K., Jensen, K.H. and Jorgensen, E. (2010) The risk of tail biting in relation to level of tail-docking. In: Lidfors, L., Blokhuis, H. and Keeling, L. (eds) *Proceedings of the 44th Congress of the International Society of Applied Ethology (ISAE), Coping in Large Groups*. Wageningen Academic Publishers, Wageningen, the Netherlands, p. 91.

Tom, E.M., Rushen, J., Duncan, I.J.H. and de Passillé, A.M. (2002a) Behavioural, health and cortisol responses of young calves to tail docking using a rubber ring or docking iron. *Canadian Journal of Animal Science* 82, 1–9.

Tom, E.M., Duncan, I.J.H., Widowski, T.M., Bateman, K.G. and Leslie, K.E. (2002b) Effects of tail docking using a rubber ring with or without anesthetic on behavior and production of lactating cows. *Journal of Dairy Science* 85, 2257–2265.

Tucker, C.B., Fraser, D. and Weary, D.M. (2001) Tail docking dairy cattle: effects on cow cleanliness and udder health. *Journal of Dairy Science* 84, 84–87.

Wada, N., Hori, H. and Tokuriki, M. (1993) Electromyographic and kinematic studies of tail movements in dogs during treadmill locomotion. *Journal of Morphology* 217, 105–113.

Wansbrough, R.K. (1996) Cosmetic tail docking of dogs. *Australian Veterinary Journal* 74, 59–63.

Watkins, L. and Maier, S. (2002) Beyond neurons: evidence that immune and glial cells contribute to pathological pain states. *Physiological Reviews* 82, 981–1011.

Watts, J.E. and Marchant, R.S. (1977) The effects of diarrohea, tail length, and sex on the incidence of breech strike in modified mulesed Merino sheep. *Australian Veterinary Journal* 53, 118–123.

Webb Ware, J.K., Vizard, A.L. and Lean, G.R. (2000) Effects of tail amputation and treatment with an albendazole controlled-release capsule on the health and productivity of prime lambs. *Australian Veterinary Journal* 78, 838–842.

Yeates, J.W., Rocklinsberg, H. and Gjerris, M. (2011) Is welfare all that matters? A discussion of what should be included in policy-making regarding animals. *Animal Welfare* 20, 423–432.

3

Fat Companions: Understanding the Welfare Effects of Obesity in Cats and Dogs

Peter Sandøe,[1*] Sandra Corr[2] and Clare Palmer[3]
[1]*University of Copenhagen, Denmark;* [2]*University of Nottingham, UK;*
[3]*Texas A&M University, USA*

3.1 Abstract

Overfeeding is arguably the most significant feeding-related welfare problem for companion animals in developed countries. From an animal welfare perspective, the amount fed presents a dilemma between avoiding feelings of hunger and maximizing animal health. This chapter will not only focus on the effects of overfeeding on the welfare of the affected animals but also on how the problem is brought about via animals' companionship with humans. Based on studies of the relationship between humans and their companion animals, a number of social and psychological factors can be uncovered that contribute to dogs and cats becoming overweight or obese. These factors may be viewed as barriers to preventing dogs and cats from maintaining a body weight that does not compromise their health and welfare. In the final section of the chapter, the importance of overcoming these barriers is discussed and ways of tackling such barriers are suggested.

3.2 Introduction

One of the most basic concerns of animal care is to ensure that animals are properly fed. This is reflected in the first principle of the 'Five Freedoms', set up to define the basic requirements of animal welfare by the British Farm Animal

*E-mail: pes@life.ku.dk

Welfare Council (now Farm Animal Welfare Committee). In its latest version, this principle is formulated so that animals should enjoy 'Freedom from hunger and thirst – by ready access to fresh water and a diet to maintain full health and vigour' (Farm Animal Welfare Committee, 2013).

Historically, underfeeding and malnutrition were major issues for dogs and cats, and images of undernourished dogs and cats figure strongly in past stories about cruelty to animals. Although less of a problem in modern times, occasionally – even in rich countries – individuals still keep large numbers of dogs and cats without being able to feed them a proper diet, while cats and dogs that lack human homes may also suffer from thirst, starvation and malnutrition.

In general, however, starvation and lack of essential nutrients are no longer major problems for dog and cat populations in richer parts of the world. Most people have sufficient resources to buy enough food for themselves and their animals. Companion animals are also viewed increasingly as family members, clearly benefiting from the affluence of their owners. Parallel to – and partly caused by – this development, there has been a significant change in the way people feed their companion animals. Increasing numbers of dogs and cats in the richer parts of the world are fed ready-made feed, either in a dry or a wet (canned) form. The production of this feed is regulated to ensure the right composition of macronutrients and a proper content of vitamins, minerals and other micronutrients (Nestle and Nesheim, 2010).

However, these diets can be fed in too large quantities. Overfeeding, particularly in combination with too little exercise, leads to obesity and increases the risk of related diseases, including cardiovascular disease, diabetes, skeletal disease and various forms of cancer. These diseases not only shorten the expected lifespan of the animals but also may have adverse effects on their quality of life. There are also likely to be negative impacts on the animals' owners in terms of worry and grief about their companions' health and early death – as well as the cost of veterinary treatment.

This chapter will focus on the problem of overfeeding dogs and cats and how the problem is brought about via the animals' companionship with humans. To a significant degree, human lifestyles influence and affect companion dogs and cats. The human population in many parts of the world suffers from an epidemic of obesity, caused by (among other things) a lifestyle with too much food, containing too much fat and sugar, accompanied by too little exercise. Companion animals, for better or worse, share this lifestyle with their owners.

To set the context, we shall begin by considering how 'fatness' is defined in dogs and cats and then look at how their health and welfare is affected when they exceed their ideal weight. This opens up the dilemma about whether maximizing longevity and minimizing morbidity, or avoiding the feeling of hunger, is better for animal welfare. Then we shall review what is known about why owners of dogs and cats allow their companions to become overweight. This review uncovers a number of barriers that stand in the way of dogs and cats remaining close to their ideal weight. We shall conclude with a discussion of various means of overcoming these barriers.

3.3 How is 'Fatness' Defined and Measured in Cats and Dogs, and How Many Animals are Affected?

In companion animals, as in humans, a distinction is drawn between being overweight and being obese. Being overweight can be defined as having a body condition where levels of body fat exceed what is considered optimal. Obesity can be defined as being overweight to the extent that serious effects on the individual's health become likely.

Various values are given in the literature for optimal percentage body fat, ranging from 15–20% for cats and dogs (Toll *et al.*, 2010) to 20–30% body fat for cats (Bjørnvad *et al.*, 2011). According to one expert in the field, the answer depends on a number of factors, including how the measurements are made and the age, breed and gender of the animal. He states that in his experience, the optimal percentage of body fat for cats is between 10 and 20%, whereas for dogs it varies between 10 and 35%, depending on breed and circumstances (Alex German, personal communication).

As more body fat translates into a higher body weight, it is possible to use body weight as a proxy measure of body fat. One recent review paper defines overweight cats and dogs as those that are 10–20% above optimal weight and obese animals as being more than 20% over optimal weight (Toll *et al.*, 2010). In another review paper, obesity is defined as occurring at 30% above ideal body weight (Burkholder and Toll, 2000). However, as noted by an expert in the field, 'these criteria have not been confirmed with rigorous epidemiologic studies, and limited data exist on the nature of an optimal body weight' (German, 2006, p. 1940).

Although most people would probably claim that they would 'know' a fat dog or a fat cat when they see one, difficulties arise in distinguishing between an animal that is overweight in comparison to one of normal weight and between an animal that is overweight and one that is obese. Thus, as well as getting to grips with definitions of overweight and obesity, it is also necessary to develop a reasonably accurate, but practical, method of assessment. In the case of humans, a simple and practical measure widely used to assess relative weight – although it is somewhat problematic and controversial – is the body mass index (BMI). Using the BMI, an individual's body fat can be estimated based on information about the person's weight and height, and used to determine whether or not the individual is of a healthy size. Such a simple measure is not easily transferable to dogs, however, as there are many diverse breeds with very different body conformations. This is less problematic in the case of cats.

However, practical measures are available to score the amount of body fat in dogs and cats. These Body Condition Scores divide dogs or cats into a number of categories, ranging from 'emaciated' to 'severely obese', based on a subjective assessment of specific features. These features include, for example, the shape of the animal viewed from above and how easily palpable the ribs are (Laflamme, 1997; McGreevy *et al.*, 2005; Toll *et al.*, 2010). Studies have shown that such measures correlate well with more advanced measurements of the amount of body fat and that, in general, there is good agreement between measurements across different users (German *et al.*, 2006).

However, there are exceptions. For example, a recent Danish study (Bjørnvad *et al.*, 2011) of indoor-confined adult neutered cats found that the body condition score tended to underestimate the level of body fat in those cats measured by means of a DEXA scanner. Due to lack of exercise, these cats are in a condition that, in the human case, has been labelled 'skinny fat'. As with some physically inactive people, due to a decrease in lean body mass, these cats have a relatively high level of body fat despite what appears to be a healthy body weight; and as in the human case, 'skinny fat' can lead to Type 2 diabetes and other serious health problems.

A number of studies have been undertaken in North America, Europe and Australia to find out what proportion of dogs and cats are overweight and obese. Most of the studies have been on the dog population and they report the prevalence of overweight and obese dogs to be between 22 and 44% (Mason, 1970; Edney and Smith, 1986; Hand *et al.*, 1989; Crane, 1991; Kronfeld *et al.*, 1991; Robertson, 2003; McGreevy *et al.*, 2005).

Two studies (Lund *et al.*, 2005, 2006) used a cross-sectional design to measure the prevalence of overweight and obese animals among a large number of cats (n = 8159) and dogs (n = 21,754) seen by US veterinarians. They found that 28.7% of adult cats were overweight and a further 6.4% were obese; for adult dogs, the numbers were 29.0% and 5.1%.

As these numbers indicate, there are significant differences in the proportion of overweight and obese cats and dogs reported in the available studies. This may, in part, reflect differences in sampling and in who has been asked (owners or vets). However, these differences may also reflect local variations. Such local variations are clearly found in the case of human obesity, where the level of obesity among adults in the USA is over 30%, compared to between 8 and 25% in European countries and around 5% in some Asian countries like Japan (World Health Organization, 2012).

Nevertheless, to make a very rough generalization, around one-third of adult cats and dogs kept as companions in the rich parts of the world are overweight, and more than one in 20 is obese. There is no evidence as to whether the proportion of overweight and obese companions is on the increase, but experts in the field seem to think that the problem is growing (German, 2006).

3.4 Is This a Welfare Problem?

A huge body of literature in veterinary medicine documents that obesity in dogs and cats increases the risk of a number of health problems (Fig. 3.1). According to one review paper, these problems include 'orthopaedic disease, diabetes mellitus, abnormalities in circulating lipid profiles, cardio-respiratory disease, urinary disorders, reproductive disorders, neoplasia (mammary tumours, transitional cell carcinoma), dermatological diseases, and anaesthetic complications' (German, 2006, p. 1940S). These conditions not only shorten the expected lifespan of the affected animals but may also lead to a reduction in their quality of life. Thus, obesity in cats and dogs can potentially lead to serious health problems that will cause them suffering (this term is used in the weak sense – see Weary, Chapter 11, this volume).

Even being moderately overweight seems to have a negative effect on animal health. It has been documented in rodents that animals fed *ad libitum* have a shorter lifespan than animals on a restricted diet (Hubert *et al.*, 2000). One experiment has been done to see whether this also holds true for dogs (Kealy *et al.*, 2002; Lawler *et al.*, 2005, 2008). This concerned two randomly selected

Fig. 3.1. This 3-year-old, 60-kg dog suffers from a ruptured cruciate ligament in one leg. As a result, he needs to rest after even very short walks. Used with permission.

groups of Labrador retriever dogs, with 24 dogs in each group. These two groups of dogs were treated in a similar way, apart from their feeding regime. One group was initially fed *ad libitum*; then feed was reduced to a level at which the dogs stayed overweight but did not become obese (with a mean body condition score around 6.5 on a scale ranging from 1, emaciated, to 9, severely obese). The other group was fed 25% less than the first group throughout the study. Dogs in the latter group remained leaner and lived longer; the median lifespan of the leaner group was 13.0 years compared to 11.2 years for the moderately overweight group. In addition, where the dogs in the leaner group developed the same diseases as those in the overweight group, the onset of disease came later and the diseases were less severe.

Thus, it appears that companion animals lose potential years of life when they are overweight. This is despite the fact that the lifespan of dogs and cats seems to have been increasing overall (Kulick, 2009), most likely due to improvements in the quality of nutrition and better veterinary care. So, feeding companion animals too much may result in suffering and premature death.

On the other hand, work with rodents and pigs indicates that a restrictive diet of the sort that secures maximum longevity and minimum morbidity causes welfare problems in the form of increased hunger and derived effects in terms of increased aggression, elevated levels of stress and the development of stereotypies (D'Eath *et al.*, 2009; Kasanen *et al.*, 2010). Assuming the same conclusions apply to dogs and cats, there may be a real dilemma here between two of the five freedoms: it may not always be possible to secure both freedom from hunger and freedom from disease.

D'Eath *et al.* (2009) have argued that this is a real dilemma in respect to keeping laboratory, companion and some farm animals. Kasanen *et al.* (2010, p. 40) have furthermore argued that in some cases the animals should be allowed free access to feed (feeding *ad libitum*, or AL): 'One could argue that the best solution is simply to proceed with AL, allowing the rats to be fat and friendly rather than lean and mean. A similar line of thought seems to apply to fattening pigs.' This solution usually works for farm and laboratory animals because they will be killed at a relatively young age, so they will never live to experience the possible negative effects later in life of having been fed too much when younger. However, this does not apply to cats and dogs that normally live until they either die naturally or are euthanized because of disease or old age.

Of course, there may be ways to make the dilemma between preventing hunger and protecting health less stark. With more exercise, animals may be able to eat more without detrimental effects to their health (and are also less likely to be skinny fat). Animals may also engage more in feeding-related behaviours without getting too much food if they are required to work for their food. Another way of dealing with the dilemma is to feed bulk diets that are less dense in calories but may provide a feeling of satiety. Nevertheless, these methods are not likely to remove fully the problem about the best feeding strategy to achieve optimal welfare.

The answer to the question about optimal feeding strategy seems, in fact, to depend on how welfare is defined. If welfare is defined in terms of fitness and function, then it may be optimal to choose a restricted feeding regime. If, on the other hand, welfare is defined as getting the maximum amount of pleasure and the minimum amount of frustration or pain – either in total over a full life or per year lived – then a more liberal feeding regime may be preferable, because a potential gain in avoidance of disease should be balanced against avoiding frustration and stress relating to hunger.

However, many dogs and cats end up becoming so fat that they do not have good welfare in terms of fitness and function, and it is also unlikely that they have good subjective welfare, because the experienced negative health effects of obesity come to outweigh the satisfaction gained from eating.

This takes us on to the obvious next question: to understand why owners who generally seem to love their cats and dogs allow these animals to grow fat, and in doing so, compromise their welfare.

3.5 Why Do Owners Allow their Companion Animals to Become Fat?

There are a number of things that owners can do to prevent cats and dogs from becoming overweight. Most importantly, they can feed them less in terms of volume and/or they can feed them diets that are less dense in calories. They may be able to exercise them more (or, in the case of cats, make more opportunities for exercise available). Yet many owners find this very difficult to do. Some reasons for this have been identified in the growing body of scientific literature on the causes of canine and feline obesity. In particular, a number of studies have investigated whether there is a link between particular characteristics of the owners of cats and dogs and the risk of their companion animals becoming obese.

First, these studies found that there was a relationship between obesity in dogs kept as companions and obesity in their owners: if the owner was overweight, it was more likely that the dog also would be obese (Mason, 1970; Kienzle *et al.*, 1998; Colliard *et al.*, 2006; Nijland *et al.*, 2010). However, the same relationship is not found between owners and their cats (Nijland *et al.*, 2010).

Second, studies seem to indicate that there is a link between the owner's income and whether the companion is overweight or obese: the poorer the owner, the more likely it is that the companion weighs too much (Kienzle *et al.*, 1998; Courcier *et al.*, 2010).

These first two findings reflect the well-documented but complex relationship between obesity and low social status in humans (Paeratakul *et al.*, 2002; Robert and Reither, 2004; Bove and Olson, 2006; McLaren, 2007), and it is not surprising that habits and constraints relating to people's own lives spill over into how they view and treat the animals in their care.

Another risk factor for being overweight and obese in both cats and dogs is neutering. Even though it is clearly possible, through careful feeding, to keep neutered cats and dogs at their ideal weight (provided that, in the case of outdoor cats, they do not have access to other households that also feed them), in practice, neutered dogs and cats are much more likely to be overweight or obese than intact ones (Nguyen *et al.*, 2004; Lund *et al.*, 2006).

So far, social, demographic and physiological factors that seem to increase the risk of an owner having an overweight companion animal have been considered. Now, a closer look will be taken at the possible psychological mechanisms that may be involved.

Many people with overweight or obese companions do not recognize that this is the case: a number of studies show that owners of overweight or obese dogs (Rohlf *et al.*, 2010; White *et al.*, 2011) and cats (Allan *et al.*, 2000; Kienzle and Bergler, 2006) underestimate the body condition of their animals. A similar situation is seen with people: parents of overweight children systematically underestimate their children's weight (Etelson *et al.*, 2003).

In the case of dog owners, this is illustrated by Fig. 3.2, where owners' estimation of the weight of their dogs is compared to body condition scores

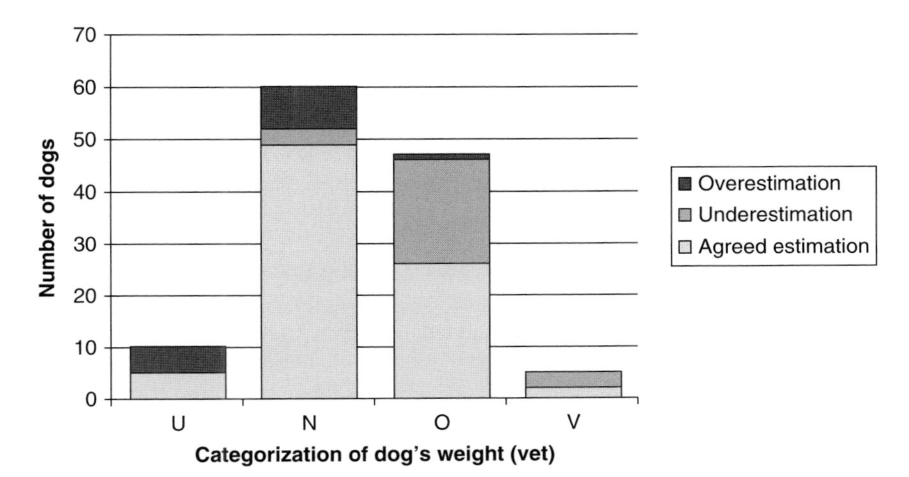

Fig. 3.2. Comparison between owner's estimation of their dog's weight and vet-assessed body condition score (BCS). Underweight (U, BCS = 2–2.5); normal weight (N, BCS = 3); overweight (O, BCS = 3.5–4.5); very overweight (V, BCS = 5). Bars are divided by owner perception types (agreed estimation, underestimation and overestimation). Owners agreed with vets for a significantly lower proportion of the overweight dogs than they did for the dogs the vet classed as normal weight (chi-square = 7.96; $P = 0.005$). (From White *et al.*, 2011. Printed with permission.)

assessed by veterinarians. Assuming that the assessments made by the veterinarians are reasonably accurate, then it follows that owners tend to underestimate the weight of overweight and very overweight dogs (and, interestingly, they also seem to overestimate the weight of underweight dogs).

Another factor seems, in the words of Shearer (2010, p. 2), to be 'linked to the dynamic of the human–animal bond. In other words, owners of overweight or obese cats and dogs see food as a convenient and acceptable form of communication and interaction with their pets.'

This line of thinking has been elaborated in two German studies. Based on a questionnaire study, Kienzle *et al.* (1998) report that there are some important differences in the relationship between owners and obese dogs in contrast to owners and normal weight dogs. Obese dogs slept more often in the owners' beds, the owners spoke more and on a greater variety of subjects to their dogs and they were less afraid of contracting diseases from their dogs than owners of dogs of normal weight. Owners of obese dogs also saw the animals as less important in relation to traditional dog roles such as sources of exercise, work and protection. The authors interpreted these differences as indicators of what they call 'overhumanizing' (Kienzle *et al.*, 1998, p. 2779): 'The obese dogs were indulged as "fellow-humans" and they were no longer treated as typical companion animals.'

Another important finding of this study was that obese dogs were more often present and fed titbits when the owner prepared or ate their own meals. The authors conclude 'that owners of obese dogs tend to interpret their dog's every need as a request for food' (Kienzle *et al.*, 1998, p. 2780).

This study also confirmed the finding that owners of obese dogs tended to be obese themselves and found that owners of obese dogs took little interest in their own health compared to owners of normal dogs. Therefore, the authors concluded, part of the explanation of the behaviour of owners of obese dogs was a transfer of the owners' eating habits and attitude to health to their dogs (Kienzle *et al.*, 1998).

A follow-up study of cats by the same German group found a number of similarities between the human factors involved in dogs and in cats becoming overweight. The similarities concerned the tendency to humanize the animals and communicate through food. Thirty per cent of owners of overweight cats, compared with 12% of owners of cats of a normal weight, stated that they did not feel very happy prior to acquiring a cat and the cat was intended to console and encourage them. According to the authors 'these results are suggestive of 1) a closer relationship between overweight cats and their owners than between normal cats and their owners, 2) more humanization of overweight cats than of normal cats, 3) a potential role of overweight cats as a substitute for human companions' (Kienzle and Bergler, 2006, p. 1948).

Another interesting finding in the case of overweight cats had to do with the sex of the owner. The relative number of female owners was higher in the overweight (97%) than in the normal weight cat groups (87%). According to

Kienzle and Bergler (2006), this may be linked to the results of other studies showing that women tend to have closer relationships with their cats than men do (Bergler, 1989).

The study also showed a difference between the owners of the overweight cats compared to the owners of the obese dogs in the earlier study. Whereas owners of obese dogs tended to be more obese, to care less about their own health and were in general less well off than owners of normal dogs, these differences were not found for cat owners (cf. also Nijl *et al.*, 2010).

The authors suggest as part of the explanation of this finding that people who are not themselves overweight or obese easily find other ways of engaging with dogs than through food, for example by taking them on walks; however, people are less aware of alternative communication modes with cats, making feeding the main contact point between many cats and their owners. That owners of overweight cats tend to communicate with their cats via feeding is further supported by the finding that they are more likely to serve the cat homemade food and to make food available *ad libitum*, and are less prone to play with their cats (Kienzle and Bergler, 2006). It should be added that all the cats in the study were indoor cats and this, of course, limited opportunities for exercise.

A final observation – which may be relevant in explaining why even owners that have no problems controlling their own weight seem to allow their cats to become overweight – is that 'cat owners were less aware of their pet's weight problem than dog owners' (Kienzle and Bergler, 2006, p. 1949). There may be different reasons for this: cats conceal clinical signs of illness more than dogs; the lifestyle of cats, involving a lot of sleeping, may contribute to making their weight less obvious; and cats appear less often in public alongside their owners than dogs and are therefore less prone to attract comments from other people. Also, cats that are allowed to go outdoors are more likely to supplement their feeding in various ways.

Some may object to these attempts to explain overweight and obese dogs and cats by reference to a 'spillover' of the owners' attitudes towards themselves. While a person's desire to eat may, for various reasons, be very strong, it is less clear how such desires may lead an individual to overfeed his or her companion animal. After all, a person experiences his or her own food cravings, but he or she does not experience the cravings of the dog or the cat.

One response to this is to note that cats and dogs may manipulate the owner. Thus, it is likely that many owners overfeed their companions because they are open to the signals the animals send. This raises an interesting analogy to the relationship that people have with their children, especially those too young to care for themselves. As in the case of companion animals, it may be argued that parents do not actually feel their child's hunger pangs but even though small children cannot open fridges, they soon learn how to make their parents do it for them. Similarly, dogs and cats probably learn how to manipulate their owners to feed them (though we do not need to suppose that this is a conscious, worked-out strategy on the part of the animal).

Problems concerning overweight and obese companions may, in many ways, be compared to problems concerning childhood obesity. Just as one would like to think that very few parents knowingly and voluntarily allow their children to become obese, it is likely that only very few owners of dogs and cats knowingly and voluntarily put the welfare and health of their companions at risk by overfeeding them. Rather, it seems that owners of fat companions are drawn into overfeeding their animals by a number of factors, some of which they have limited control over, much in the same way as they are drawn into overfeeding themselves and their children.

So far, this chapter has discussed work linking the demographic and psychological characteristics of cat and dog owners with the risk of the animal becoming overweight and obese. The findings are based on a number of studies conducted in different countries and paint a coherent picture of a number of factors that increase the likelihood of dogs and cats becoming overweight or obese. However, the total number of studies on the social and psychological mechanisms underlying the way people feed their companion animals is still small, and conclusions will probably develop as new research becomes available.

The next step, then, must be to discuss how to deal with these factors.

3.6 Whether and How to Prevent and Treat Problems with Overweight Companion Animals

As should be clear from the previous section, it is not all that easy for owners of cats and dogs to prevent their companions from becoming overweight, especially if the owner is highly connected to his or her animal, has strong empathetic responses to it and regards food as a primary means of communication with it. So, there are a number of difficulties that must be overcome for owners of cats and dogs to control their companions' weight.

One difficulty is that a companion animal being overweight or obese does not matter to the same degree to all owners. There may be several reasons for this. Most basically, some people may lack the financial or personal resources even to handle issues relating to their own health, so the long-term health of their companion animals may not be high on their agenda.

But we also know that the issue of weight control gives rise to controversies in the case of humans: some people and groups resist the idea that it is inevitably problematic for someone to be overweight or obese. For example, in the USA, various movements have developed in favour of so-called 'fat acceptance' (Kirkland, 2008). Followers of these movements deliberately term themselves fat and argue that size is no more a reason for discrimination or prejudice than height, sex or skin colour. They resist the medicalization of being fat and try to show that fat people are just normal people in large bodies who mostly are perfectly capable of living good lives and making a contribution by work and otherwise. In light of this, one might expect similar attitudes to develop in relation to companion animal obesity.

Also, there are perspectives that take a critical view on the utilitarian-like welfare maximization approach that might be thought to permeate this chapter. Of course, from all main ethical points of view it should be granted that, *other things being equal*, it is a good thing that companion animals stay healthy and, to the extent that keeping down their weight is a means to maintain their health, it is a good thing to keep down their weight. However, according to this view, other things are not always equal and therefore there may, in certain circumstances, be reasons to downplay efforts to maintain a normal weight of the companion animals in one's care.

For example, on some kinds of relational view (see Palmer and Sandøe, 2011, for an overview of different ethical views and references to further literature), the key concern in our dealings with companion animals is to maintain and nurture close, caring relationships. Here, it may turn out that always focusing on limiting food intake could be detrimental to maintaining such a caring relation between the human owner and the companion dog or cat. Of course, it will never be acceptable for someone with a caring attitude to allow a companion dog or cat to become morbidly obese. However, in line with the thinking of the fat acceptance movement, being overweight to some degree could be accepted as the lesser of two evils – where the greater evil is that the concern to restrict food disrupts the caring bond between the owner and his or her companion animal.

To control the weight of cats and dogs, it is essential from the beginning of the animal's life to impose a balanced feeding regime, to check the animal's weight regularly and, if necessary, to adjust feeding. Regular exercise (in the case of dogs) and other activities initiated by the owner will also help to prevent the animals becoming overweight. Particularly in the case of cats, enabling owners to find ways of engaging with their animals that are not linked to feeding, for example through various forms of training and playing, could break the link between contact and feeding. It is also important to consider whether or not to neuter the animal; leaving the animal intact is one measure that should reduce the animal's likelihood of becoming overweight (Palmer *et al.*, 2012).

As already discussed, many people do not recognize that their companion animals are overweight or obese and, even if they do, seem unable to manage their animal's weight. Veterinarians therefore have an important role in making owners aware of the problem when the animals are brought to the surgery for vaccination or regular health checks.

When people become aware of their cat or dog having gained too much weight, they may seek advice on how to deal with it. A number of approaches have been developed, mainly through collaboration between pet food companies and veterinarians.

First, as mentioned above, tools have been developed that allow owners to assess the relative weight of their cat or dog. By means of the body condition score, owners can find out and regularly check whether or not their companion is a healthy size.

Second, a number of products have been developed to help owners of cats and dogs to control their animals' weight. Some feed products are low in calories and high in fibre, comparable to 'light' products for human consumption; others are high in protein. These products may, if used appropriately, be helpful as part of either a weight-loss or a weight-maintenance plan. Such feed products may, in some cases, help to deal with the previously discussed dilemma between satisfying hunger and avoiding the health threats of being overweight. Other products that may enable engaged owners to deal with the same issue are feeding devices that allow the animal to spend more time engaged in feeding-related behaviours, without getting more food.

Third, weight-loss programmes have been developed which provide guidance for owners of overweight dogs or cats to do something about the problem (Toll *et al.*, 2010). These programmes involve feeding controlled diets and regular weighing of the animal, often combined with exercise and enrichment regimes. Persevering with such a programme can be challenging – it takes many months before most animals will start to lose weight, even on a special diet, and owners must remain motivated for months to see any effects. A key element of such a programme, therefore, is regular engagement with staff at a veterinary or special weight-control clinic. There are also various guidebooks on the market (e.g. Harvey and Taylor, 2012) and Internet-based programmes, typically sponsored or run by pet food manufacturers (e.g. PetFit.com).

Fourth, weight-maintenance plans have been developed that enable an owner to keep his or her companion animal at an optimal weight, once the desired weight loss has been achieved (Toll *et al.*, 2010). As in the case of human obesity, the difficulty is not only the initial weight loss but also the long-term change of lifestyle required to maintain the weight at a healthy level.

Finally, a drug for dogs, Slentrol (Pfizer), is available in the USA. This down-regulates a dog's appetite, so that even if an owner is unable to control how much he or she feeds the dog, the drug still makes the dog want to eat less. For some, this may be the easiest solution. As it says on the part of the product's web page directed at dog owners:

> We know you care about your dog's health. Yet, helping him or her lose unhealthy weight can be difficult. Diet and exercise are often hard to stick with and may not be enough. If you're frustrated with traditional weight-loss methods, now there's a new option – Slentrol.
>
> (Pfizer, 2013)

However, the use of such drugs may have adverse side effects: Pfizer's own Slentrol web page notes that 'the most common side effect is vomiting, especially during the first month of treatment and when the dose is increased'. Given the risk of such side effects, and that the dog's welfare is the purpose of the treatment, Slentrol should be viewed as the last resort when other less potentially unpleasant approaches have been shown not to work and where using the drug is likely, on balance, to produce welfare benefits.

A striking feature of many of these solutions is the size of commercial vested interests involved. In contrast to human nutrition, where a significant part of the research is undertaken by independent researchers funded by government grants and where independent institutions advise the public, nearly all research and advice on companion animal nutrition is sponsored and run by producers of pet food and pharmaceuticals. The dominance of vested interests does not necessarily imply that the research in the field and the published advice are skewed, but it may give rise to distrust. This, in turn, may give rise to lack of trust in the solutions offered; a quick look at the many web pages and self-help books promoting alternative ways of feeding dogs and cats certainly reveals a widespread distrust in the methods offered by the established players in the pet food market.

Another striking feature of the proposed solutions is that they all appeal to individual action and do not focus on the political, social and related human issues that seem to underlie the problem. An important prerequisite for people to be able to appreciate the problems related to the overfeeding of their companions may be that they are able to understand and deal with their own unhealthy lifestyle. Tackling these social issues will require solutions to problems far greater than the issue of obesity of companion animals.

Finally, given these social issues, as well as owners' psychological states, it is important to avoid moralizing, i.e. condemning people for doing what they cannot help doing – to some degree, at least. There have been cases, in Sweden and the UK (Kulick, 2009), where obese pets have been forcibly removed from their owners by the authorities. While this may be a necessary step when the welfare of the cat or dog in question is severely threatened, forced removal of the animal and punishment of the owner cannot be the general solution to the widespread problems concerning obesity in cats and dogs. Condemning owners of fat companions without either helping them, assisting them to help themselves or dealing with underlying social problems may contribute to the widespread difficulties of stigmatization and low self-esteem that already face obese people in many societies (Puhl and Brownell, 2006). Instead, solutions must be identified to solve the underlying problem, which is to enable owners of cats and dogs to develop a healthy lifestyle for themselves and their companions.

3.7 Conclusions

- In the past, insufficient feeding was a widespread problem among companion dogs and cats. Now, in the richer parts of the world, the biggest problem in relation to feeding companion animals is overfeeding: about one in three dogs and cats kept as a companion in developed countries is overweight and about one in twenty is obese.
- Simple tools have been developed to differentiate between normal, overweight and obese companion animals. There is uncertainty regarding how exactly these divisions link to problems concerning the health and welfare of the animals affected, but it is likely that being even moderately overweight results in diminished life expectancy, and obesity will often lead to serious disease and related grave welfare problems.
- A restricted diet may also have a negative effect on animal welfare. When it comes to feeding companion animals, there seems to be a real dilemma between protecting the animal against hunger and protecting its long-term health.
- Overweight and low-income owners are more likely to have an overweight dog, suggesting that this issue is indicative of broader social problems. It seems that owners are also more likely to overfeed animals they view as being like fellow humans or with which they have close emotional ties.
- Preventing companion animals from getting fat and trying to slim companion animals that have become overweight or obese is a good thing, viewed from most ethical points of view. However, there may be a discussion to be had about the priority of this compared to other concerns – for example, the concern to uphold a harmonious relationship between owner and animal.
- A number of potential solutions are available to owners who want to prevent or treat the problems associated with overweight companion animals.

Questions for discussion

1. Your neighbour's pet is overweight. Should you point this out?
2. Why is it difficult to define an optimal body weight for a cat or a dog?
3. How does one's definition of animal welfare affect how one views the problem of dogs and cats becoming overweight or obese?
4. Why may people disagree about the need to control the weight of dogs and cats?
5. How should the problem of growing levels of obesity among dogs and cats be dealt with by our society, and how is this linked to dealing with problems concerning human obesity?

Acknowledgements

The authors gratefully acknowledge useful comments provided on earlier versions of this chapter by Mike Appleby, Charlotte Reinhard Bjørnvad, Lise Lotte Christensen, Tove Christensen, Sandra Edwards, Björn Forkman, Ayoe Hoff, Vibeke Knudsen, Jesper Lassen, Morten Ebbe Juul Nielsen, Helle Friis Proschowsky, Signild Vallgårda, Dan Weary and the other contributers to this volume. Special thanks are due to Alex German, who inspired the first author of this chapter to start thinking about cat and dog obesity and who helped sort out some key technical issues.

References

Allan, F.J., Pfeiffer, D.U., Jones, B.R., Esslemont, D.H. and Wiseman, M.S. (2000) A cross-sectional study of risk factors for obesity in cats in New Zealand. *Preventive Veterinary Medicine* 46(3), 183–196.

Bergler, R. (1989) *Man and Cat: The Benefits of Cat Ownership*.Blackwell Scientific Publications, London.

Bjørnvad, C.R., Nielsen, D.H., Armstrong, J., McEvoy, F., Hølmkjær, K.M., Jensen, K.S., *et al.* (2011) Evaluation of a nine-point body condition scoring system in physically inactive pet cats. *American Journal of Veterinary Research* 72(4), 433–437.

Bove, C.F. and Olson, C.M. (2006) Obesity in low-income rural women: qualitative insights about physical activity and eating patterns. *Women and Health* 44(1), 57–78.

Burkholder, W.J. and Toll, P.W. (2000) Obesity. In: Hand, M.S., Thatcher, C.D., Remillard, R.L. and Roudebush, P. (eds) *Small Animal Clinical Nutrition*. Mark Morris Institute, Topeka, Kansas, pp. 404–406.

Colliard, L., Ancel, J., Benet, J.J. and Paragon, B.M. (2006) Risk factors for obesity in dogs in France. *Journal of Nutrition* 136, 1951–1954.

Courcier, E.A., Thomson, R.M., Mellor, D.J. and Yam, P.S. (2010) An epidemiological study of environmental factors associated with canine obesity. *Journal of Small Animal Practice* 51(7), 362–367.

Crane, S.W. (1991) Occurrence and management of obesity in companion animals. *Journal of Small Animal Practice* 32, 275–282.

D'Eath, R.B., Tolkamp, B.J., Kyriazakis, I. and Lawrence, A.B. (2009) 'Freedom from hunger' and preventing obesity: the animal welfare implications of reducing food quantity or quality. *Animal Behaviour* 77(2), 275–288.

Edney, A.T.B. and Smith, P.M. (1986) Study of obesity in dogs visiting veterinary practices in the United Kingdom. *Veterinary Record* 118, 391–396.

Etelson, D., Brand, D.A., Patrick, P.A. and Shirali, A. (2003) Childhood obesity: do parents recognize this health risk? *Obesity Research* 11(11), 1362–1368.

Farm Animal Welfare Committee (2013) http://www.defra.gov.uk/fawc/about/five-freedoms/ (accessed 21 January 2013).

German, A.J. (2006) The growing problem of obesity in dogs and cats. *Journal of Nutrition* 136, 1940–1946.

German, A.J., Holden, S.L., Moxham, G.L., Holmes, K.L., Hackett, R.M. and Rawlings, J.M. (2006) A simple reliable tool for owners to assess the body condition of their dog or cat. *Journal of Nutrition* 136, 2031–2033.

Hand, M.S., Armstrong, P.J. and Allen, T.A. (1989) Obesity: occurrence, treatment, and prevention. *Veterinary Clinics of North America: Small Animal Practice* 19, 447–474.

Harvey, A. and Taylor, S. (2012) *Caring For An Overweight Cat.* Vet Professionals, Roslin, Midlothian, Scotland.

Hubert, M., Laroque, P., Gillet, J. and Keenan, K.P. (2000) The effects of diet, *ad libitum* feeding, and moderate and severe dietary restriction on body weight, survival, clinical pathology parameters, and cause of death in control Sprague-Dawley rats. *Toxicological Sciences* 58(1), 195–207.

Kasanen, I.H.E., Sørensen, D.B., Forkman, B. and Sandøe, P. (2010) Ethics of feeding – the omnivore dilemma. *Animal Welfare* 19(1), 37–44.

Kealy, R.D., Lawler, D.F., Ballam, J.M., Mantz, S.L., Biery, D.N., Greeley, E.H., *et al.* (2002) Effects of diet restriction on lifespan and age-related changes in dogs. *Journal of the American Veterinary Medical Association* 220(9), 1315–1320.

Kienzle, E. and Bergler, R. (2006) Human–animal relationship of owners of normal and overweight cats. *Journal of Nutrition* 136, 1947–1950.

Kienzle, E., Bergler, R. and Mandernach, A. (1998) A comparison of the feeding behavior and the human–animal relationship in owners of normal and obese dogs. *Journal of Nutrition* 128, 2779–2782.

Kirkland, A. (2008) Think of the hippopotamus: rights consciousness in the fat acceptance movement. *Law and Society Review* 42(2), 397–431.

Kronfeld, D.S., Donoghue, S. and Glickman, L.T. (1991) Body condition and energy intakes of dogs in a referral teaching hospital. *Journal of Nutrition* 121, 157–158.

Kulick, D. (2009) Fat pets. In: Tomrley, C. and Naylor, A.K. (eds) *Fat Studies in the UK*. Raw Nerve Books, York, UK, pp. 35–50.

Laflamme, D.P. (1997) Development and validation of a body condition score system for dogs. *Canine Practice* 22, 10–15.

Lawler, D.F., Evans, R.H., Larson, B.T., Spitznagel, E.L., Ellersieck, M.R. and Kealy, R.D. (2005) Influence of lifetime food restriction on causes, time, and predictors of death in dogs. *Journal of the American Veterinary Association* 226(2), 225–231.

Lawler, D.F., Larson, B.T., Ballam, J.M., Smith, G.K., Biery D.N., Evans, R.H., *et al.* (2008) Diet restriction and aging in the dog: major observation over two decades. *British Journal of Nutrition* 99(4), 793–805.

Lund, E.M., Armstrong, P.J., Kirk, C.A. and Klausner, J.S. (2005) Prevalence and risk factors for obesity in adult cats from private US veterinary practices. *International Journal of Applied Veterinary Medicine* 3(2), 4–6.

Lund, E.M., Armstrong, P.J., Kirk, C.A. and Klausner, J.S. (2006) Prevalence and risk factors for obesity in adult dogs from private US veterinary practices. *International Journal of Applied Veterinary Medicine* 4(2), 3–5.

McGreevy, P.D., Thomson, P.C., Pride, C., Fawcett, A., Grassi, T. and Jones, B. (2005) Prevalence of obesity in dogs examined by Australian veterinary practices and the risk factors involved. *Veterinary Record* 156, 695–702.

McLaren, L. (2007) Socioeconomic status and obesity. *Epidemiologic Reviews* 29, 29–48.

Mason, E. (1970) Obesity in pet dogs. *Veterinary Record* 86, 612–616.

Nestle, M. and Nesheim, M.C. (2010) *Feed Your Pet Right – The Authoritative Guide to Feeding Your Dog and Cat.* Free Press, New York, London, Toronto and Sydney.

Nguyen, P.G., Dumon, H.J., Siliart, B.S., Martin, L.J., Sergheraert, R. and Biourge, V.C. (2004) Effects of dietary fat and energy on body weight and composition after gonadectomy in cats. *American Journal of Veterinary Research* 65, 1708–1713.

Nijland, M.L., Stam, F. and Seidell, J.C. (2010) Overweight in dogs, but not in cats, is related to overweight in their owners. *Public Health Nutrition* 13(1), 102–106.

Paeratakul, S., Lovejoy, J.C., Ryan, D.H. and Bray, G.A. (2002) The relation of gender, race and socioeconomic status to obesity and obesity comorbidities in a sample of US adults. *International Journal of Obesity* 26, 1205–1210.

Palmer, C. and Sandøe, P. (2011) Animal ethics. In: Appleby, M.C., Mench, J.A., Olsson, I.A.S. and Hughes, B.O. (eds) *Animal Welfare, 2nd edition*. CAB International, Wallingford, UK, pp. 1–12.

Palmer, C., Corr, S. and Sandøe, P. (2012) Inconvenient desires – should we routinely neuter companion animals? *Anthrozoos* 25S, 153–172.

Pfizer (2013) https://www.slentrol.com/display.aspx (accessed 18 January 2013).

Puhl, R.M. and Brownell, K.D. (2006) Confronting and coping with weight stigma: an investigation of overweight and obese adults. *Obesity* 14, 1802–1815.

Robert, S.A. and Reither, E.N. (2004) A multilevel analysis of race, community disadvantage, and body mass index among adults in the US. *Social Science and Medicine* 59(12), 2421–2434.

Robertson, I.D. (2003) The association of exercise, diet and other factors with owner-perceived obesity in privately owned dogs from metropolitan Perth, WA. *Preventive Veterinary Medicine* 58(1–2), 75–83.

Rohlf, V.I., Toukhsati, S., Coleman, G.J. and Bennett, P.C. (2010) Dog obesity: can dog caregivers' (owners') feeding and exercise intentions and behaviors be predicted from attitudes? *Journal of Applied Animal Welfare Science* 13(3), 213–236.

Shearer, P. (2010) *Literature Review – Canine, Feline and Human Overweight and Obesity*. Banfield Applied Research and Knowledge Team (BARK), Portland, Oregon, pp. 1–8.

Toll, P.W., Yamka, R.M., Schoenherr, W.D. and Hand, M.S. (2010) Obesity. In: Hand, M.S., Thatcher, C.D., Remillard, R.L., Roudebusch, P. and Novotny, B.J. (eds) *Small Animal Clinical Nutrition*. Mark Morris Institute, Topeka, Kansas, pp. 501–544.

White, G.A., Hobson-West, P., Cobb, K., Craigon, J., Hammond, R. and Millar, K.M. (2011) Canine obesity: is there a difference between veterinarian and owner perception? *Journal of Small Animal Practice* 52(12), 622–626.

World Health Organization (2012) *World Health Statistics 2012*. WHO, Geneva, Switzerland.

Welfare and Quantity of Life

4

Nuno H. Franco,[1] Manuel Magalhães-Sant'Ana[1,2] and I. Anna S. Olsson[1]*
[1]*IBMC, Universidade do Porto, Portugal;* [2]*Escola Universitária Vasco da Gama, Portugal*

4.1 Abstract

Animal welfare science is focused mostly on evaluating and improving the quality of life of animals that actually exist. This leaves out a range of ethically relevant issues regarding the quantity of life – in terms of number of animals living and the longevity of each animal. In many cases, quantity and quality are related, and often there is a tension between the two. In this chapter, we develop a discussion around four practical cases presenting quality/quantity dilemmas: (i) the issue of dairy cow longevity; (ii) the early slaughter of male dairy calves; (iii) the killing of newly hatched male layer chicks; and (iv) the conflict between reduction and refinement in animal research. The practical, economic and animal welfare aspects characterizing each case are presented, together with relevant stakeholders' perspectives. We discuss the cases in light of the most relevant currents of thought in animal ethics, highlighting the main values at stake and which possible solutions may be sought according to each perspective.

4.2 Introduction

Most of the attention in animal welfare science, be it in practical research or in theoretical discussion, is given to how animals live their lives; that is, the

*E-mail: olsson@ibmc.up.pt. The first two authors contributed equally to this chapter.

quality of life. But dilemmas around the killing of animals also involve considerations of the value of life. This kind of discussion is perhaps most visible in companion animal medicine, where euthanasia may relieve a severely ill animal from further suffering but at the same time breaks a strong human–animal bond and leaves a grieving owner alone. It is the companion animal angle that Sandøe and Christiansen (2007) take as a starting point for their analysis of the ethical issues at stake. However, related dilemmas arise in all fields of animal use. This chapter relies on cases from farming and animal experimentation to widen the discussion.

In particular, when moving outside of the companion animal field, the concept of quantity takes on two possible meanings: quantity as lifespan (longevity) of individual animals and quantity as the number of animals (at a given moment in time or accumulated numbers over time). Both of these aspects are relevant for a discussion on the value of animal life in farm and laboratory animals. This chapter considers them both in discussing the following four dilemma cases:

1. Dairy cow longevity where concern is arising over the decreasing lifespan of dairy cows, which seems to present a choice between more cows living less time (and possibly with worse welfare) or fewer cows living more time (and supposedly a better life).
2. Male dairy calves which may be slaughtered at less than a week of age (thus having extremely short lives) or fattened for a few months under questionable conditions to produce veal (longer lives but of debatable quality).
3. Male layer chicks that are typically killed at hatching and where the most plausible alternative to be developed seems to be a further shortening of this extremely short existence.
4. Laboratory animals where the reuse of animals in multiple experiments and the principle of reduction of animal numbers give rise to dilemmas between quality of life and quantity of animals.

These dilemmas involve issues having to do with the quality and duration of the lives of animals that will be born and live, and the discussion will focus on these issues. However, some issues of whether or not certain animals should be brought into existence will also be addressed.

There are some more fundamental philosophical issues that are relevant to life and death decisions affecting animals but which go beyond the scope of this chapter. These include the challenge of deciding whether the life of another individual of a different species is worth living, and questions having to do with whether animals have a will to live or a notion of the future and how having or lacking these concepts interacts with the harm caused by death. Others (e.g. Yeates, 2011; Bruijnis *et al.*, 2013) have addressed these questions in detail and the interested reader is referred to these texts.

To shed light on the four situations, we consider the practical issues involved – such as the existence and feasibility of alternatives – in combination with the ethical issues at stake. We develop the discussion against the background of the major theoretical considerations around balancing quality and quantity, which implies considering the value of animal life, and also in the light of relevant surveys and focus group discussions.

4.3 Four Practical Cases

4.3.1 Longevity of dairy cows

There is widespread concern over a decrease in the longevity of dairy cattle (although reversing trends have been reported (Farm Animal Welfare Council, 2009)). Before the intensification of farming practices during the second half of the 20th century, there were records of Jersey cattle living more than 25 years (Odlum, 1950), but today it is common for Holstein cows to be culled at 4 years of age. The reasons for this dramatic decrease in life expectancy are multiple. It has been argued that they derive from the genetic selection of dairy cows for increased milk yield: high-producing dairy cows suffer from production diseases such as lameness, mastitis, ketosis and reduced fertility (Oltenacu and Broom, 2010) (and may also have other welfare problems: Fig. 4.1). These diseases have great impact on the health and the welfare of animals, and on their productivity. It is often the low economic viability that supports the decision of (early) culling. In a recent analysis of cattle mortality in France between 2003 and 2009 (and relying on a database of 75 million animals), Perrin and colleagues (2011) identified a peak of mortality in both beef and dairy cows at around 3 years of age. In beef cattle, killing the animal is a requirement to obtain the product. In contrast, milk is produced by living animals and the mortality peak at 3 years requires a different explanation; however, the cited study does not distinguish between different reasons for mortality. There is also a complex interaction between genetic progress, the availability of replacement heifers and decision making over culling. A farmer may keep most or even all healthy heifers to ensure sufficient replacement, and for each heifer ready to calve, the least productive cow in the herd will be culled. Considering genetic progress, the heifer can be assumed to be 'genetically superior' to the cow she replaces: so at the time, culling and replacement will be a sensible decision, even though in a larger perspective the decline in longevity driven by such decisions may be questionable (Erling Strandberg, personal communication).

High-yield systems have a somewhat contradictory effect on the number of existing lactating cows: fewer animals are needed to produce the same amount of milk but there is an increased need of replacement heifers to renew the short-lived animals. In Europe, the total volume of milk production has

Fig. 4.1. Holstein-Friesian dairy cow conformation has changed over years of selective breeding, resulting in taller, longer and thinner cows. Note that the lying cow in this photo from a UK dairy farm is longer than the bedded stall. Photo: Manuel Magalhães-Sant'Ana.

been constant for several decades, despite a gradual decline in the number of dairy cows that exist at any given time (Hocquette and Chatellier, 2011), a trend that can also be found in other developed countries (FAOSTAT, 2012). This has other consequences outside the dairy industry because, after slaughter, dairy cows provide meat and leather. The dairy sector represents 57% of global cattle meat production (FAO, 2010), although this figure includes bull calves of dairy breeds raised specifically for meat production.

Although duration of life is rarely considered an animal welfare issue, longevity can be used as an indicator of welfare. Bruijnis and co-workers use the case of dairy cow lameness as a proxy to consider longevity as a constitutive element of animal welfare. They argue that an animal should be allowed to live long enough to have the opportunity to perform species-specific behaviours and to flourish; an important part of 'natural living', a concept they include in animal welfare (Bruijnis *et al.*, 2013). Natural living is a concept particularly pertinent within the context of organic farming, but Vonne Lund does not seem to include longevity as an element of natural living in her seminal paper (Lund, 2006). As shall be discussed later in this chapter, from an ethical perspective it makes a difference whether increased longevity means a longer life with the same health and welfare issues that (now, at least) partly motivate early culling, or a long life as a consequence of an improved health status.

The dairy cow longevity question and the male dairy calf question, discussed below, are related. Increasing cow longevity reduces the need for replacement heifers. This may make it economically viable for farmers to inseminate cows with the most desirable phenotype with sexed dairy breed semen to obtain replacement heifers and use semen from beef breeds for the remaining cows. With this approach, only those calves purposely bred to become replacement heifers would be of a full dairy type, whereas those that would go into meat production would be cross-bred. However, the price for sexed semen is roughly double that of normal semen, and many cows require multiple inseminations (reduced fertility being a production-related disorder). Therefore, as long as most female calves born are considered to be needed as replacement heifers, sexed semen is unlikely to be the preferred option and the problem of male calves of a dairy genotype is likely to persist.

4.3.2 The male dairy calf

With the increasing specialization of cattle breeds into dairy and meat production, the value of rearing dairy calves for meat production has decreased. This phenomenon has long been known in breeds with a traditional pronounced dairy phenotype such as the Jersey, but is now increasingly also affecting the originally more dual-purpose Holstein-Friesian cattle. The practical consequences of this have not been taken to such extremes as in poultry production (see below), as male Holstein-Friesian calves are still being raised and slaughtered, but at least in some countries a considerable proportion of calves are killed during the first week of age, as raising them for later slaughter is not considered economical.

Contrary to many of the ethical issues in animal production, the birth of unwanted bull calves of dairy breeds is not a problem exclusive to large-scale modern intensive animal production. To produce milk, a dairy cow must give birth to a calf about once every year. Despite the changes in milk yield, growth rate and food turnover, 'one lactation = one calf' has remained unchanged since the beginning of dairy farming. To generate the same amount of milk from cows with a lower milk yield, more calves would have to be born. In her book, *Animal Machines* (1964), Ruth Harrison pointed out the problematic consequences of this dependency and gave figures for her era (in which the average UK dairy cow produced approximately half the amount of milk as that produced by an early 21st century cow (Oltenacu and Broom, 2010)):

> An unavoidable characteristic of the rearing of animals is that approximately the same number of male and female offspring will be produced. It follows that where cows are kept for milk there is the problem of what to do

> with male calves, many of which are not suitable for rearing as beef because the strain has been developed primarily for its milking potential. It has been estimated that the surplus of unwanted calves, 'bobby calves', in this country amounts to some 800,000 to 1,000,000 yearly.
>
> (Harrison, 1964, p. 62)

Not very different from 50 years ago, these calves go down one of three possible routes. They may be slaughtered as bob(by) calves, at only a few days of age, or they may be fattened for slaughter for the production of veal (up to 8 months of age, according to European marketing regulations) or for beef/young beef/rose veal. To produce the characteristic light-coloured meat, veal calves are raised mainly on milk products, whereas calves slaughtered at an older age are fed a cereal-based diet (AVMA, 2008; Sans and de Fontguyon, 2009).

Of the estimated 30 million dairy calves born in 2008 in EU-27, just under 6 million (i.e. over 40% of male dairy calves) went into veal calf production. In Europe, this production is greatly dominated by two countries, the Netherlands and France. In other European countries, male dairy calves that are not slaughtered shortly after birth are either fattened until 8–12 months of age or exported to one of the veal production countries (Sans and de Fontguyon, 2009). In the USA, in terms of animal numbers, veal production is split fairly equally between calves slaughtered a few days after birth and calves formula-fed until 16–20 weeks of age, with a slight majority of bobby calves (e.g. USDA, 2011).

Veal calf production is controversial because it is associated with a range of animal welfare issues, starting with the transport and commingling of week-old calves from different dairy farms and followed by the specific rearing conditions and diet designed to produce the typical white veal meat. Whereas the most extreme housing conditions in which calves are crated or tethered in permanent darkness are now outlawed in Europe (European Union, 1997), single housing and tethering throughout the fattening period is still widespread in the USA (AVMA, 2008). Across the world, veal calf production implies prolonging the period that calves are fed mainly liquid food (milk replacement) beyond what would be physiologically normal and maintaining the animals with a low iron status. Even though current European legislation requires a minimum provision of fibrous food, a recent epidemiological study has shown that poor rumen development and abomasal lesions are still frequent in calves from European white veal production (Brscic *et al.*, 2011)

Whereas the issue at stake in veal calf production is primarily feeding and housing conditions, slaughtering calves shortly after birth is controversial on the grounds of the extremely short life these animals are given. In combination, these issues motivated the RSPCA and Compassion in World Farming to convene a stakeholder forum to discuss actions to change the future of UK dairy calves. The early 21st century UK situation very clearly illustrated several aspects of the dilemma. As described by Ruth Harrison, there is a tradition of exporting bobby calves for veal production in continental Europe. During the years when this market was unavailable as a result of the BSE-motivated export ban on UK cattle, on-farm killing of male newborn calves increased, to the point that it had practically doubled by 2006, the year in which the ban was lifted. Nevertheless, even then an estimated 138,700 of the 438,000 male dairy calves born in the UK were killed on the farm (Beyond Calf Exports Stakeholder Forum, 2008). Such early on-farm killing of calves is usually combined with destruction of the carcass; thus, the meat is not used for human consumption. Contrary to practice in Europe, the main route (91%) for Australian non-replacement dairy calves is commercial slaughter at 5–7 days: an estimated 700,000 calves a year are slaughtered this way (Animal Health Australia, 2011).

4.3.3 The male layer chick

Today's poultry production is highly specialized, to the extent that all commercial production is dominated by lines genetically selected for meat or egg laying presented by a few multinational companies. The sex of the birds is an issue for production considerations in both meat and egg production, but while a female broiler chicken is still useful, albeit slightly less productive than her male counterpart, male layer chicks are of no commercial value; they do not lay eggs and their slender bodies and slow growth make them unable to compete with broilers for meat production (even though this has been tried).[1] Day-old male chicks of layer lines are presently killed, either by exposure to CO_2 gas (after which the carcasses can be used as animal feed) or by maceration (instant death but more limited use of the carcasses; see Leenstra *et al.*, 2011, for a discussion of these alternatives).

Leenstra and collaborators (2011) investigated the view of Dutch citizens on how to manage male layer chicks in poultry production, using a combination of focus group interviews and an Internet-based survey. Participants were asked to choose between and comment on ten alternative approaches divided into three main groups (Table 4.1).

The participants were unaware of the practice of killing day-old male chicks and were initially shocked to learn about it. When discussing the issue and the list of alternatives in focus groups, people considered a number of aspects, including not only animal friendliness, naturalness, risks for human and animal safety but also practical considerations such as feasibility, as well as resource and financial economics. No clear preferred option was evident, but 'the study indicated that most people would support the pursuit of technological alternatives', with the preferred technological alternatives being 'i) looking into the fresh egg (to determine the sex of the egg and not incubate male eggs); ii) influencing the laying hens such that they produce fewer male

Table 4.1. Potential ways to manage the problem of male chicks of layer breeds which have no commercial or production value and are currently killed at 1 day old. (From Leenstra *et al.*, 2011.)

Technological solutions
Looking into the egg
1. Determining the sex of freshly laid eggs and not incubating male eggs
2. Determining the sex of early embryos and destroying the male embryos
3. Determining the sex of late embryos and destroying the male embryos
Changing the hen
4. Environmentally influencing the hens to produce fewer male eggs
5. Crossing the parents in such a way that male embryos are not viable
Genetic modification
6. To facilitate sexing of freshly laid eggs and not incubating male eggs
7. To make sex reversal of male embryos into female chickens possible
8. Such that male embryos die during early development
Other solutions
9. Accepting the current practice of killing day-old chickens
10. Less specialized chickens, so that the males can be used for meat production (dual-purpose chickens)

eggs; and iii) using genetic modification to facilitate sexing fresh eggs'. Participants were also in favour of the idea of a dual-purpose type chicken, even though they recognized that this was not a very realistic option.

Of the alternatives presented by Leenstra and co-workers, only one is actually being considered for practical use: examining samples of incubating embryos in order to destroy male embryos. It is not yet in commercial use, but there seem to be only relatively minor practical obstacles left before it would be feasible to implement this practice (Michael Clinton, personal communication, May 2012).

4.3.4 Reducing, reusing and refining in animal experimentation

The use of animals for scientific purposes raises specific issues in regards to the value of life versus the quality of life. The restrictive conditions under which laboratory animals are housed are not worse (and are sometimes better) than those in which production animals live, but the fact that animals in biomedical research are often intended to model disease presents particular challenges when it comes to providing 'freedom from pain, injury or disease'. Therefore, it is sometimes not possible to provide laboratory animals with 'a life worth living' (see Yeates, 2011, for an overview of the origin of the concept). In such situations, early killing may be the most effective way to relieve welfare problems.

Fortunately, in most cases, there are effective measures to reduce such distress and improve the well-being of animals used in research ('refinement', one of the three R's proposed by Russell and Burch, 1959) without compromising research results (Fig. 4.2). Some such refinements imply sharing the burden by several animals, so that each animal is exposed to less accumulated distress. This is sometimes described as the 'fairness to the individual' approach (Tannenbaum, 1999), but it conflicts with another of the three R's, 'reduction' (of animal numbers), as improving the well-being of each individual animal is done at the cost of using more animals (de Boo *et al.*, 2005; Olsson *et al.*, 2012). The practical situation in which this dilemma is most evident is in the choice between reusing animals from a previous experiment or using new animals. Another case where refinement and reduction collide is the choice between group housing and single housing of animals in experiments requiring data at the cage level, such as in dietary research (Festing and Altman, 2002). Group housing of social animals is a refinement, but in this case the cage becomes the experimental unit and group housing will hence result in more animals used.

We presented such reduction–refinement dilemmas to 195 participants in eight laboratory animal science courses held in four different institutions; most (83% overall) would rather house mice in pairs than individually (Franco

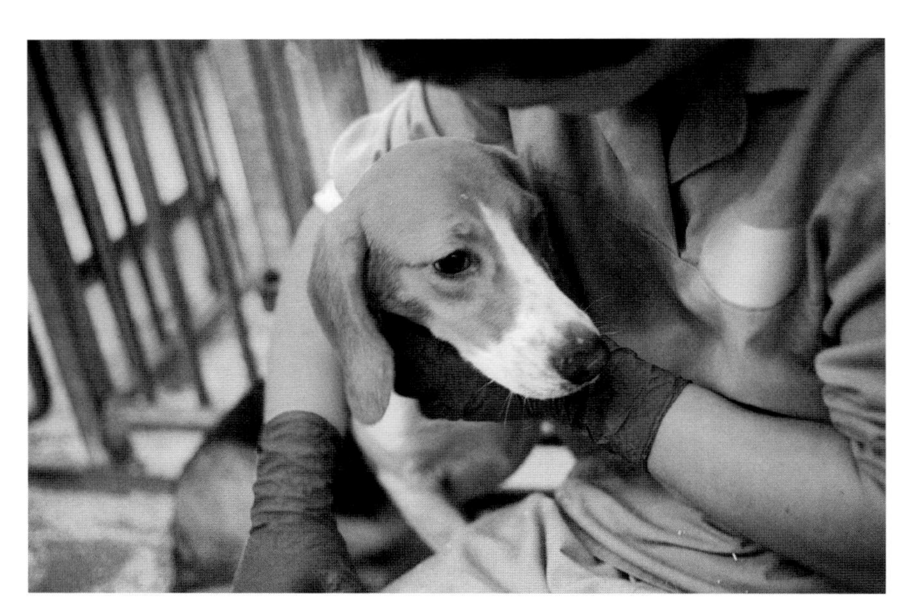

Fig. 4.2. Repeated use of animals is not always in conflict with animal welfare. Dogs used in pharmacokinetic studies in which low doses of drugs are administered are an example of how research animals can be given long and healthy lives within the research setting. Photo: Understanding Animal Research.

and Olsson, 2013). Biology undergraduate students (n = 71) with no experience in using animals in research responded in a similar way, with 85% favouring refinement and 15% favouring reduction (own unpublished data).

A more drastic reduction–refinement dilemma was presented to participants in 11 laboratory animal science training courses (n = 235), who were asked to choose between two hypothetically equally valid approaches for a given experiment: (i) conducting 20 considerably painful procedures – but with no permanent damage – on the same mouse across 20 days or (ii) conducting one single procedure per mouse on 20 mice. In both options, animals would be euthanized at the end of the experiments. This could be a realistic albeit drastic representation of the reuse situation. Answers to this dilemma were more evenly distributed (49% for using 20 mice; 51% for 20 trials with the same mouse). When asked what their approach would be for other species, regardless of their previous answer, almost every respondent chose the same approach for rabbits (99% of those who had chosen 20 trials on one mouse and 91% of those who had chosen to use 20 animals). However, when it came to dogs and primates, many of those who had previously opted for using 20 mice changed their approach, namely 31% for dogs, 21% for rhesus macaques and 38% for chimpanzees (32, 56 and 48% for biology undergraduate students). For these respondents, the way one values the life of a given animal versus the quality of life for each animal appears to depend on which species the animal belongs to. Using large numbers of animals to avoid cumulative suffering is seen as a more acceptable approach for mice and rabbits than for primates and dogs. This was further substantiated by answers to a question about whether animals should be adopted or moved to sanctuaries after the experiment; rehabilitation was considered most important for companion animal species and non-human primates (Fig. 4.3).

4.4 Philosophical Perspectives on Practical Cases

After having introduced the central issues regarding the four case studies, this chapter moves to an appraisal of each one in light of the animal ethics theories of more relevance for these cases. These include utilitarianism, animal rights theory, virtue ethics, relational perspective, contractarianism, Judeo-Christian morals and environmental ethics.

Utilitarian theory relies on the aggregate consequences of actions, i.e. the right action is the one that produces the best overall good. While hedonistic utilitarians like Bentham and Mill held that we should act to maximize net happiness (Singer, 2011), Peter Singer (2011, p. 13) proposed that our actions should aim to do what on balance 'furthers the interests of those affected'. In the utilitarian take on animal ethics, the capacity to experience suffering and pleasure is usually taken as the basis of interest (Singer, 1975). While quality of life certainly is central to utilitarian theory, that this line of thinking does

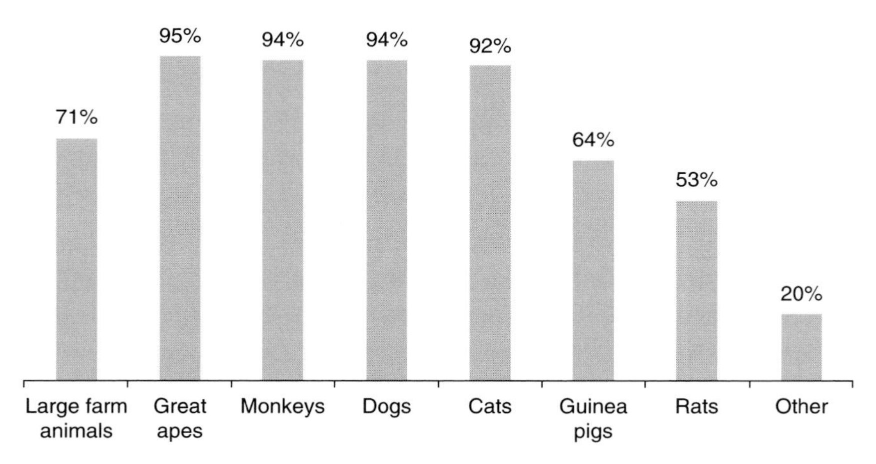

Fig. 4.3. Distribution of answers to the question: 'Animals used in experiments are sometimes transferred to "sanctuaries", or even given for adoption – instead of being euthanized – when their research purpose ends and rehabilitation is possible. However, that is not the case for most species. In your opinion, for which animal species should this kind of measure be considered, whenever possible?' (From Franco and Olsson, 2013.)

not consider each individual life as valuable in itself has consequences for its view on killing as well as for bringing individuals into existence. Therefore, the short lifespan of dairy cows can be seen as posing little ethical concern in itself as long as the life of one cow is replaced by the life of another milk-producing cow. In this way, the life of a sentient being is replaced by the life of an equivalent one and the end result is no more or no less pleasure or suffering than if the same cow had been maintained for a longer life. Or, in fact, there may even be a net gain if the cow to be culled has health problems which imply pain and suffering, such as lameness or mastitis (typical production-related diseases in dairy cows), whereas the heifer to replace her is healthy. However, this is not to say that a dairy production system in which cows last for one or two lactations only is always unproblematic in utilitarian terms. If health and welfare problems develop as the result of how cows are bred, fed or managed, this is clearly an issue of utilitarian relevance. A similar argument can be made in the case of male dairy calves. The consumer demand for white veal could (at least theoretically)[2] be satisfied either through slaughtering a greater number of dairy calves at a very early age (and consequently low weight), as in Australia, or by rearing these calves for longer, as in Europe. And given the particular welfare problems associated with veal production, it may actually, from the utilitarian perspective, be better for a calf to live a very short but relatively painless life than to have to undergo first transport, often for long distances, and then rearing with an inappropriate diet. Interestingly, the proposed solution to avoid killing of day-old male layer chicks may actually remove this case from the realm of utilitarian moral preoccupation, at least if the arguments of Mellor and Diesch (2007) are accepted that chicks have very limited or even no conscious neural activity before hatching. As for the refinement–reduction conflict in laboratory animals, there is really no dilemma for the utilitarian; a greater number of animals living a less distressful life is clearly preferable to a smaller number of animals enduring more individual suffering.

In contrast to the utilitarian outlook, the value of animal life is central to *animal rights theory.* This approach is based on an extension of the Kantian concept of intrinsic value to all sentient beings, a view that inherently affords animals the right to be treated always as ends in themselves (Regan, 1989). Built into this right is the right to life, and therefore killing an animal is only acceptable under some very special circumstances: under extreme situations of self-defence, when two rights to life stand against each other or when being killed is in the rights-holder's own interest. This view presents what appears to be the most compelling argument for favouring long lives. However, the animal rights view would *a priori* reject both farming and experimentation on the grounds that animals are ends in themselves and cannot be used as means to an end. That is, from the animal rights perspective, maintaining a production animal is unacceptable, independently of its longevity. Still, having no dairy cows works for future cows but fails to address the lives of the existing ones; in order to preserve their rights, these animals would have to be housed in sanctuaries, with minimum human interference until the moment they die from old age. Interestingly, this is what a recent UK initiative to produce Ahimsa, or slaughter-free, milk proposes to do (www.ahimsamilk.org).

A second set of theories focuses primarily on the role of the human actor rather than on the animal as the subject of the act. *Virtue ethics* proposes that moral conduct is based on the practice of fundamental virtues and the avoiding of vices, essentially seen from an anthropocentric perspective. Through a *relational perspective* on ethics, the moral status of animals is affected by the type of relationship humans establish with these (types of) animals. According to a relational point of view, we have much stronger moral responsibility towards animals in our care or to which we can relate closely than towards others (Sandøe and Christiansen, 2008; Palmer and Sandøe, 2011).

Good animal husbandry is essential in a *virtue ethics perspective.* A virtuous, responsible farmer provides his or her cows with the best possible care. The relational aspects are particularly important: animal husbandry implies a bond between the farmer and the animals. This bond is usually greater the longer it endures. In practical terms, this means that a compassionate farmer would only break the bond for the benefit of the cows (e.g. enduring suffering) or because of the prospect of economic loss. The research animal issue may lead to very different conclusions, starting from a virtue ethics point of view, depending on one's overall view of the value of animal research. For a person who seriously questions that animal research is useful, supporting it would not be virtuous because if no good comes out of the research, it must be considered unnecessary harm (Hursthouse, 2006). If one sees animal research as a sound means to develop therapies to cure or ease the burden of disease, then it becomes a moral imperative to carry on using animals, albeit as humanely as possible. The killing of animals used in experiments may be seen as a compassionate attitude if this is the best measure to avoid unacceptable suffering or if the scientific objectives are already met and the animals cannot be rehabilitated for reuse or rehoming. On the other hand, whenever the conditions so permit, allowing animals to go on living a life worth living may be seen as the virtuous thing to do.

The *relational perspective* gives humans moral duties primarily (or even exclusively) towards animals in human care or to which humans can in any way relate closely (like primates) but much less so (or not at all) towards others (Sandøe and Christiansen, 2008; Palmer and Sandøe, 2011). This explains why only 18% of European citizens disapprove of the use of mice in biomedical research but 37% disapprove of the use of dogs and primates for the same purpose (Crettaz von Roten, 2012). This differentiation seems to have consequences also for determining the value of life – at least that seems a plausible interpretation of why researchers consider non-human primates and companion animals much more appropriate for rehabilitation (thus avoiding killing) than rodents. As the relational perspective also gives value to the human–animal relationship, it would also support keeping the same dairy cow for longer rather than replacing her. It is unlikely that such a relationship exists for newly hatched chicks or newborn dairy calves.

One influential approach going back to Hobbes is *contract theory*. In this, and in more contemporary versions of contractarianism such as defended by Jan Narveson and others, animals have no moral standing. The underlying idea of contractarian ethics is that moral obligations derive from the mutual agreement between different parties. As animals can neither claim their rights nor demand any duties from humans, the classical contractarian perspective on animals is that any animal's life and welfare are relevant only in so far as they matter to other humans. However, some animal ethicists have described the human relationship with farm animals in terms of a contract (Lund *et al.*, 2004; Rollin, 2008) in which the animals are actually partners. Such a tacit partnership could also fit into a contemporary Christian perspective on animals. The *Judeo-Christian tradition* assigns humans dominion over creation (from Genesis 1:26–30), but there are different interpretations of how this dominion should be exerted (Fellenz, 2007). The despotic perspective rests on the assumption that nature, including animals, was created by God to serve humans, whereas the stewardship perspective emphasizes the human responsibility to take care of creation (Barad-Andrade, 1991). The official standing of the Catholic Church today is much closer to the stewardship view, in stating that 'animals, by their mere existence ... bless [man] and give him glory. Thus men owe them kindness' (The Holy See, 1993, Part Three, Section Two,

Chapter Two, Article 7:2416). Moreover, although the use and killing of animals for food, clothing and research is seen as legitimate – and therefore not sinful – the church clearly points out that this must be done within limits and in such a way as to avoid unnecessary suffering (The Holy See, 1993). The best option in a quantity/quality dilemma will again depend on the situation; these theories give no standard recommendations either way.

A concrete example of how a classical *contractarian* attitude may affect the issue of longevity is patent in recital 26 of the 63/2010/EU directive, which states that 'animals such as dogs and cats should be allowed to be rehomed in families as there is a high level of public concern as to the fate of such animals' (European Commission, 2010). Thus, to meet the implicitly contracted obligations towards citizens and voters, governmental agencies grant special protection to specific groups of animals, in virtue of their value in the eyes of the public. As regards dairy farming, the acceptable lifespan of the animal is a direct result of its productivity, from a classic contractarian point of view. If increasing the turnover of their livestock is in the interests of farmers, then that is the right thing to do. Maintaining the profit margins and obligations towards retailers and consumers are arguments used by farmers to justify farming practices with high turnovers. On the other hand, such obligations towards consumer demands can be drivers for less intensive animal farming regimes, as a result of increasing public concern for animal welfare, of which biological (organic) dairy and meat are an example. Contract theories that include animals as partners would rather favour allowing these animals to live longer, at least if their health and welfare would permit it, in a similar way as in virtue ethics (see above).

From a modern *Judeo-Christian perspective*, cruel treatment of animals is of more ethical concern than the sacrificing of animal lives. Humans are entitled to use animals, but inflicting unjustified, avoidable suffering is morally condemnable. Some factory farming practices – such as the artificial fattening of geese for *foie gras* production and battery caging of laying hens – have even been explicitly condemned by the theologian, Joseph Ratzinger, later Pope Benedict XVI (Ratzinger and Seewald, 2002). Industrializing animal production to the extent that animals are treated as mere instruments is a degradation of animals' nature and place in creation, and hence an abuse of the stewardship granted to humans. Such instrumentalization may correspond to the dairy cow case when production means that cows are regularly culled after only one or two lactations, possibly because they are no longer considered sufficiently productive to be maintained in the herd. As for male dairy calves, their birth is a natural and unavoidable consequence of dairy production, and thus slaughter of bobby calves for food is ethically preferable to rearing them under conditions which are detrimental to their welfare. Veal production, however, if conducted in such a way that animals are provided a good life, is not in itself immoral. As regards laying hens, industrialized practices for sorting and destroying male chicks may be seen as a degradation of animals to mere disposable commodities. Early identification of male embryos or, alternatively, rearing of males for food despite their low productivity seem to be more in line with modern Christian standing on the ethical treatment of animals and respect for animals' nature and purpose. Following the same rationale, the Christian view on the use of animals in biomedical research is that it is justified as long as it is conducted humanely and directed towards the benefit of human health (Pacholczyk, 2006). In that sense, animal welfare takes precedence over longevity or whatever number of animals is used, and thus our actions should favour refinement when in conflict with reduction.

Environmental ethics suggest that we should take into consideration the more global issues of nature and not only the particular aspects of the life and welfare of individuals. Such considerations rarely give a direct answer to the question of quantity versus quality, but may have indirect implications. It could be argued that the present approach of the genetic selection of cattle with strong focus on milk yield fails in respecting the nature of the cow: cows are very different from their ancestors (taller and thinner, with larger udders) and less fit to live natural lives. The loss of gene pool diversity also means that the species may have lost valuable traits needed to adapt to changes in the environment. Dual-purpose breeds and breeds with greater genetic diversity would be preferable from this perspective as regards both cattle and poultry. However, some environmental ethicists consider domesticated species so dependent on humans and detached from nature that they are more of an artefact than rightful members of the natural world (Palmer and Sandøe, 2011). From this perspective, measures to improve production efficiency may actually be an advantage if this lessens the negative environmental impact on production. Within an environmental perspective, issues such as sustainability and resource management also need to be taken into account. Animal production has a substantial impact on the global water footprint (Mekonnen and Hoekstra, 2010) and the carbon footprint (FAO, 2010; Flysjö *et al.*, 2012), and some have suggested that increasing the longevity of dairy cows could lessen the environmental footprint by reducing 'the replacement rate and the number of non-productive animals' (Boichard and Brochard, 2012, p. 548). However, an overall evaluation must also include the environmental consequences of shifting meat production from dairy to beef herds. It is unclear whether such an assessment is possible with the existing methods for assessing the environmental impact of livestock products (see de Vries and de Boer, 2010, for a review of life cycle assessments of livestock products).

4.5 Discussion

In this chapter, we have discussed four real-life cases in which there is some sort of conflict between the quality and the duration of life. We have used several philosophical theories to shed light on the issues at stake, providing various answers to what the best solution may be to the different dilemmas. Similarly, studies of public perception show a diversity of views, both among the general public and among specialists. Whereas there may be some 'wrong' answers, it is less likely that there will be a single 'right' answer to any of the dilemmas. In fact, this is exactly what characterizes a moral dilemma: there is no answer to it which does not carry its own ethical cost.

Traditionally, research and practical efforts in animal welfare have been focused on avoiding suffering, and the importance of positive experiences have really only been highlighted during the last decade (e.g. Boissy *et al.*, 2007; Yeates and Main, 2008). This is probably reflected in some bias towards anti-suffering considerations in the analysis presented in this chapter, with relevant consequences for how the quantity/quality dilemma is approached. In its most extreme form, focusing on avoiding suffering may make it 'morally right to kill off everybody to prevent them from suffering' (Sandøe and Christiansen, 2007, p. 548). In contrast, much of the moral concern over calves and chicks being killed shortly after birth/hatching is probably derived from the notion that these animals were never given the opportunity to live and to experience the good things of life.

There is also some bias towards the (more concrete) question of how to handle existing animals, as opposed to the (more abstract) question of which animals should come into existence. In many cases, the concrete question could have been avoided if the abstract question had actually been addressed. This is true for two of our four cases: with existing technology (sexed semen), dairy farmers can be sure that mostly heifers are born and technology may also (at least in the future) prevent the conception of male layer chicks. Of course, whether this distinction is at all relevant depends on whether there is a fundamental difference between terminating a life early and not allowing a life to start (cf. Bateson, 2013, for a related debate). Society certainly seems to distinguish between terminating a life and not allowing it to start if the life is human: people are generally free to choose whether to have children but are not allowed to kill each other. There are also important differences in terms of the resources needed to generate full-term young and, at least in the case of mammals, in the pain and distress associated with the birth and separation of a mother from her offspring.

4.6 Conclusions

- There are real-life cases in which there is some conflict between the quality and the duration of life.
- In dairy cows, the increase in milk production has been accompanied by an increase in production-related diseases; both longevity and quality of life are decreasing.

- Male dairy calves are of little value for rearing as beef; these calves are typically killed during the first week of life, experiencing very little life at all, or are reared for veal production under conditions of low quality of life.
- Dairy cow longevity and male dairy calves raise interconnected questions. Improved cow longevity would potentially make it economically viable for farmers to combine insemination with sexed dairy semen to generate replacement heifers, with meat breed insemination for the remaining reproduction, thus ensuring that only those calves that would become replacement heifers were of a full dairy type, whereas those that would go into meat production would be cross-bred.
- Male layer chicks have no commercial value. They are killed as soon as the sex can be determined, presently as day-old chicks. Several alternative approaches have been discussed, but the only economically realistic alternative under consideration seems to be killing at an even earlier time, i.e. before hatching.
- In research using animals, there is sometimes a dilemma between striving to reduce total numbers of animals and to reduce the impact on individual animals, highlighted by the potential to reuse animals in different procedures. Among scientists, there is no consensus of which of these principles should be given priority.

Questions for discussion

1. Imagine that you are responsible for running a zoo. A choice must be made between letting the lions have cubs every summer (and later euthanizing the surplus) or using reproductive control so that the lions only have offspring when needed to renew the population. Which policy would you prefer and why?
2. Is there something like a minimum appropriate duration of life – how long a life needs to be to be meaningful?
3. Cow longevity seems to be affected both by selective breeding (genotype) and by management practice. What can farmers do to influence the longevity of their cows? Are there other relevant actors?
4. Laboratory rodents are usually euthanized at the end of experiments, whereas more efforts are made to keep larger animals for longer periods. What are the advantages and disadvantages for the animals? What reasons are there to make a difference between species?

Acknowledgements

Thanks to Michael Clinton, Roslin Institute, for providing information on the feasibility of different technological alternatives to the killing of day-old chicks; to Eddie Bokkers, Wageningen University, for input on veal calf production and welfare; to Erling Strandberg, Swedish University of Agricultural Sciences, for input on dairy cow longevity; and to Jean-Baptiste Perrin, Institut National de la Recherche Agronomique, for the additional information on the mortality of dairy cattle. Thanks also to the editors as well as to the authors of other chapters of this book, who participated in the workshop held in Vienna in July 2012 for constructive criticism. The data on researcher attitudes come from Nuno Franco's PhD project funded by the Fundação para a Ciência e a Tecnologia (SFRH/BD/38337/2007). Manuel Sant'Ana is in receipt of a doctoral grant from Fundação para a Ciência e a Tecnologia (SFRH/BD/46879/2008).

Notes

[1]Leenstra *et al.* (2011, pp. 37–38) report that 'Experiments have been performed in a number of countries to rear layer-type males to a live weight of approximately 600 g and market them as an alternative for quail, or to a live weight of approximately 2000 g and market them as an alternative to broiler chickens ... [but they] require virtually twice the amount of feed and three times as much time to reach the required bodyweight, compared to broilers.'

[2]To test the practical feasibility would require simulations (taking into account the number of animals potentially available and considerable differences in slaughter weight of bobby calves and veal calves) that go beyond the scope of this chapter.

References

Animal Health Australia (2011) Australian Animal Welfare Standards and Guidelines – Land Transport of Livestock. Proposed Amendment to the Land Transport of Livestock Standards (SB4.5). Bobby Calves Time Of Feed Standard. Decision Regulation Impact Statement (as at 6/7/2011) Edition 1.0.

AVMA (2008) Welfare implications of the veal calf husbandry (https://www.avma.org/KB/Resources/Backgrounders/Pages/Welfare-Implications-of-the-Veal-Calf-Husbandry-Backgrounder.aspx, accessed 14 January 2013).

Barad-Andrade, J. (1991) Stewardship: whose creation is it anyway? *Between the Species* 7, 102–109.

Bateson, P. (2013) Debate: 'Is it better to have lived and lost than never to have lived at all?'. In: Wathes, C., Corr, S., May, S., McCulloch, S. and Whiting, M. (eds) *Veterinary and Animal Ethics: Proceedings of the First International Conference on Veterinary and Animal Ethics*. Wiley-Blackwell, London, pp. 286–299.

Beyond Calf Exports Stakeholder Forum (2008) Report on conclusions and recommendations (http://calfforum.rspca.org.uk/web/calfforum/reports, accessed 7 June 2013).

Boichard, D. and Brochard, M. (2012) New phenotypes for new breeding goals in dairy cattle. *Animal* 6, 544–550.

Boissy, A., Manteuffel, G., Jensen, M.B., Moe, R.O., Spruijt, B., Keeling, L.J., *et al.* (2007) Assessment of positive emotions in animals to improve their welfare. *Physiology and Behavior* 92, 375–397.

Brscic, M., Heutinck, L.F.M., Wolthuis-Fillerup, M., Stockhofe, N., Engel, B., Visser, E.K., *et al.* (2011) Prevalence of gastrointestinal disorders recorded at postmortem inspection in white veal calves and associated risk factors. *Journal of Dairy Science* 94, 853–863.

Bruijnis, M.R.N., Meijboom, F.L.B. and Stassen, E.N. (2013) Longevity as an animal welfare issue applied to the case of foot disorders in dairy cattle. *Journal of Agricultural and Environmental Ethics* 26, 191–205.

Crettaz von Roten, F. (2012) Public perceptions of animal experimentation across Europe. *Public Understanding of Science* (http://pus.sagepub.com/content/early/2012/02/01/0963662511428045.abstract, accessed 8 August 2013).

de Boo, M.J., Rennie, A.E., Buchanan-Smith, H.M. and Hendriksen, C.F.M. (2005) The interplay between replacement, reduction and refinement: considerations where the three Rs interact. *Animal Welfare* 14, 327–332.

de Vries, M. and de Boer, I.J.M. (2010) Comparing environmental impacts for livestock products: a review of life cycle assessments. *Livestock Science* 128, 1–11.

European Commission (2010) Directive 2010/63/EU of the European Parliament and of the Council of 22 September 2010 on the protection of animals used for scientific purposes. *Official Journal of the European Union* L 276 53, 33–79.

European Union (1997) Council Directive 97/2/EC of 20 January 1997 amending Directive 91/629/EEC laying down minimum standards for the protection of calves. *Official Journal of the European Communities* L 025, 24–25.

FAO (2010) *Greenhouse Gas Emissions from the Dairy Sector – A Life Cycle Assessment.* Food and Agriculture Organization of the United Nations, Rome.

FAOSTAT (2012) Trends in the livestock sector. In: *FAO Statistical Yearbook 2012.* Food and Agriculture Organization of the United Nations, Rome, pp. 198–201.

Farm Animal Welfare Council (2009) Opinion on the welfare of the dairy cow (http://www.fawc.org.uk/pdf/dcwelfar-091022.pdf, accessed May 2012).

Fellenz, M.R. (2007) Broader philosophical considerations. In: *The Moral Menagerie: Philosophy and Animal Rights.* University of Illinois Press, Illinois, pp. 33–56.

Festing, M.F. and Altman, D.G. (2002) Guidelines for the design and statistical analysis of experiments using laboratory animals. *ILAR Journal* 43, 244–258.

Flysjö, A., Cederberg, C., Henriksson, M. and Ledgard, S. (2012) The interaction between milk and beef production and emissions from land use change – critical considerations in life cycle assessment and carbon footprint studies of milk. *Journal of Cleaner Production* 28, 134–142.

Franco, N.H. and Olsson, I.A.S. (2013) Scientists and the 3Rs: attitudes to animal use in biomedical research and the effect of mandatory training in laboratory animal science. *Laboratory Animals* (http://lan.sagepub.com/content/early/2013/08/12/0023677213498717.abstract, accessed 21 August 2013).

Harrison, R. (1964) *Animal Machines.* Vincent Stuart, London.

Hocquette, J.-F. and Chatellier, V. (2011) Prospects for the European beef sector over the next 30 years. *Animal Frontiers* 1, 20–28.

Hursthouse, R. (2006) Applying virtue ethics to our treatment of the other animals. In: Welchman, J. (ed.) *The Practice of Virtue: Classic and Contemporary Readings of Virtue Ethics.* Hackett Publishing Co, Indianapolis, pp.136–154.

Leenstra, F., Munnichs, G., Beekman, V., van den Heuvel-Vromans, E., Aramyan, L. and Woelders, H. (2011) Killing day-old chicks? Public opinion regarding potential alternatives. *Animal Welfare* 20, 37–45.

Lund, V. (2006) Natural living – a precondition for animal welfare in organic farming. *Livestock Science* 100, 71–83.

Lund, V., Anthony, R. and Rocklinsberg, H. (2004) The ethical contract as a tool in organic animal husbandry. *Journal of Agricultural and Environmental Ethics* 17, 23–49.

Mekonnen, M.M. and Hoekstra, A.Y. (2010) *The Green, Blue and Grey Water Footprint of Farm Animals and Animal Products: Volume 1: Main Report.* UNESCO-IHE, Delft, the Netherlands.

Mellor, D.J. and Diesch, T.J. (2007) Birth and hatching: key events in the onset of awareness in the lamb and chick. *New Zealand Veterinary Journal* 55, 51–60.

Odlum, G. (1950) Longevity in dairy cattle. *Farmers' Weekly* 32, 54.

Olsson, I.A.S., Franco, N.H., Weary, D.M. and Sandøe, P. (2012) The 3Rs principle – mind the ethical gap! *ALTEX* 29, 333–336.

Oltenacu, P.A. and Broom, D.M. (2010) The impact of genetic selection for increased milk yield on the welfare of dairy cows. *Animal Welfare* 19, 39–49.

Pacholczyk, T. (2006) Animal rights vs human rights. *Making Sense Out of Bioethics* (http://www.ncbcenter.org/Page.aspx?pid=288, accessed 14 January 2013).

Palmer, C. and Sandøe, P. (2011) Animal ethics. In: Appleby, M.C., Mench, J.A., Olsson, I.A.S. and Hughes, B.O. (eds) *Animal Welfare. 2nd edition.* Cambridge University Press, Cambridge, UK, pp. 1–12.

Perrin, J.-B., Ducrot, C., Vinard, J.-L., Hendrikx, P. and Calavas, D. (2011) Analyse de la mortalité bovine en France de 2003 à 2009. *Inra Productions Animales* 24, 235–244.

Ratzinger, J. and Seewald, P. (2002) *God and the World: Believing and Living in Our Time: A Conversation with Peter Seewald.* Ignatius Press, San Francisco.

Regan, T. (1989) The case for animal rights. In: *Animal Rights and Human Obligations – Second Edition.*Prentice Hall, New Jersey, pp. 105–114.

Rollin, B. (2008) The ethics of agriculture: the end of true husbandry. In: Dawkins, M.S. and Bonney, R. (eds) *The Future of Animal Farming: Renewing the Ancient Contract.* Blackwell Publishing, Oxford, UK, pp. 7–20.

Russell, W.M.S. and Burch, R.L. (1959) *The Principles of Humane Experimental Technique.* Methuen and Co Ltd, London.

Sandøe, P. and Christiansen, S.B. (2007) The value of animal life: how should we balance quality against quantity? *Animal Welfare* 16, 109–115.

Sandøe, P. and Christiansen, S.B. (2008) *Ethics of Animal Use.* Blackwell Publishing, Oxford, UK.

Sans, P. and de Fontguyon, G. (2009) Veal calf industry economics. *Revue de Médecine Vétérinaire* 160, 420–424.

Singer, P. (1975) *Animal Liberation: A New Ethics for Our Treatment of Animals.* New York review: distributed by Random House, New York.

Singer, P. (2011) About ethics. In: *Practical Ethics (Third Edition)*. Cambridge University Press, Cambridge, UK, pp. 1–15.
Tannenbaum, J. (1999) Ethics and pain research in animals. *ILAR Journal* 40, 97–110.
The Holy See (1993) *Catechism of the Catholic Church*. Libreria Editrice Vaticana, Vatican City.
USDA (2011) Estimated calf slaughter under Federal inspection. In: *Annual Meat Trade Review*. Livestock and Grain Market News Service, Des Moines, pp. 12.
Yeates, J.W. (2011) Is 'a life worth living' a concept worth having? *Animal Welfare* 20, 397–406.
Yeates, J.W. and Main, D.C.J. (2008) Assessment of positive welfare: a review. *Veterinary Journal* 175, 293–300.

5

Improving Farm Animal Welfare: Is Evolution or Revolution Needed in Production Systems?

Maria José Hötzel*
Universidade Federal de Santa Catarina, Brazil

5.1 Abstract

Remarkable advances in the understanding of animal sentience, and in the development and validation of scientific methods to assess farm animal welfare, have been achieved in the past 50 years. Nonetheless, farm animal welfare improvements have been limited by interactions between economic, political, technical and biological factors. Improving farm animal welfare with moderate changes in the current production systems may become even harder as we are faced with the demand to reduce the environmental impact of livestock production and produce food for a growing, richer human population. 'Sustainable intensification' of production, which is perceived by many as the best path to reverse the environmental impacts of agriculture, is likely to exacerbate many common welfare problems. A shift to large-scale, confined production of monogastrics of highly productive genotypes may help improve feed efficiency and reduce production costs associated with energy and waste, but this shift may also cause welfare problems. Animal welfare is an important ethical social concern and needs to be integrated into the concept of sustainable agriculture. The push for further intensification ignores the fact that the public tends to reject industrial animal production systems. To achieve significant improvements in farm animal welfare, besides proposing technical innovations we need to envisage new political and economic arrangements that permit efficient and ethical animal production systems to flourish.

*E-mail: mjhotzel@cca.ufsc.br

5.2 Introduction

Animal production and productivity have increased remarkably in the last decades (Fig. 5.1), mostly as a result of intensification and industrialization of livestock systems. This process, which involved changes to housing, genotypes and nutrition with the goal of increasing production efficiency and taking advantage of economies of scale, started in industrialized countries after the Second World War. Now, more than half the world's livestock are reared in intensive operations (Naylor *et al.*, 2005; Steinfeld *et al.*, 2006).

Beginning in the 1960s in industrialized countries, members of the public concerned with the welfare of farm animals have pushed, with some success, for changes in husbandry and housing practices that they associate with intensive confined livestock production. In response to public demands, scientists have developed methods of assessing and improving farm animal welfare, with a focus on the problems identified within these industrial systems. On-farm improvements in animal welfare, however, have been limited by the incompatibility between proposed alternatives, perceived economic or practical constraints within the industry and 'welfare dilemmas' – problems that cannot be overcome without causing other similar or greater welfare problems.

Today, some other issues regarding livestock production have gained relevance for the public and policy makers. After global warming gained prominence on the global political scene, livestock agriculture has been challenged with demands to mitigate or reverse its contribution to greenhouse gas emissions, water depletion and pollution, and losses of arable soils and biodiversity.

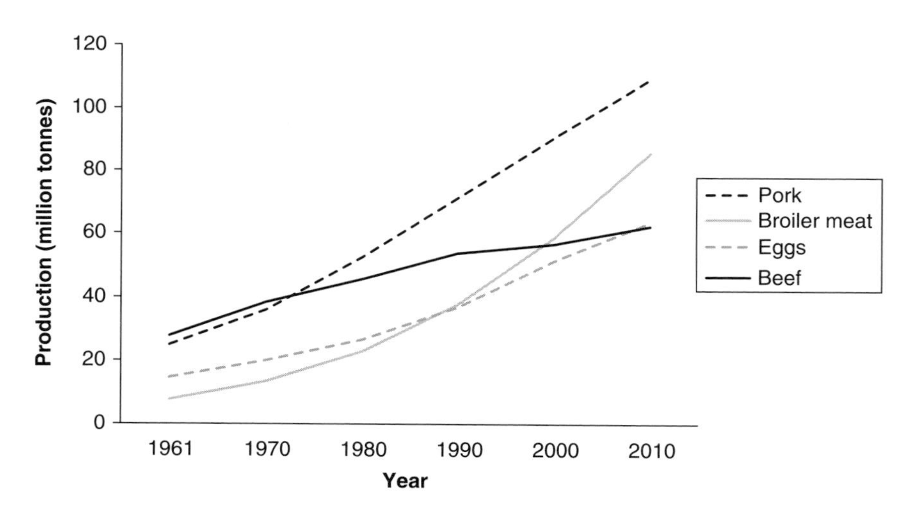

Fig. 5.1. Estimated world production of pork, broiler meat, eggs and beef from 1961 to 2010. (Source: FAO, 2011b, with permission.)

Livestock already consume about 30% of global cereals (Keyzer *et al.*, 2005), and in the near future, the demand for cereals for animal feed may exceed production capacity (Keyzer *et al.*, 2005). The increasing demand for energy that is projected to accompany the demand for food (Nonhebel and Kastner, 2011), together with the use of crops for biofuel production, has added new interrelations between livestock, energy production and food security that call for international policy reform (Naylor, 2011; Rosset, 2011). As emphasized by Naylor (2011, p. 247), 'energy markets, financial transactions, and macro-economic policies now influence agricultural markets – including both commodities and land – in unprecedented ways'. Ways to answer these challenges, either with further intensification of livestock production or via broader changes in the organization of food production to allow the (re)integration of animal, plant and soil systems, are debated intensely today in academic and political environments. An analysis of livestock production with farm animal welfare in mind needs to consider these wider challenges.

5.3 Citizens' Opposition to the Industrial Livestock Production Model

Recognition by the public of the reality of industrialization of animal production was an original driver for the farm animal welfare movement almost 50 years ago. The very use of the term 'factory farming' (Harrison, 1964) symbolizes the public rejection of the industrial intensive system, which is based on the perception that this form of farming is typically unable to provide good living conditions for farm animals (Brambell, 1965; Miele, 2010; Miele *et al.*, 2011). Over the years since the publication of Ruth Harrison's book, much of the world's population has become wealthier and more urban, and now many people have little or no contact with farms or farm animals. This disconnection explains why citizens – in industrialized countries where this has been studied – tend to overestimate the proportion of livestock that are reared in free-range and small-scale systems (Miele, 2010; Lusk and Norwood, 2011), and may also explain why the public is often sensitive to campaigns that expose the reality of livestock farming.

Today, many people perceive small-scale production in outdoor systems and natural feeding practices as more welfare friendly and better for human health and for the environment, and reject the practices they associate with the industrial systems (European Commission (EC), 2007; Miele, 2010; Naald and Cameron, 2011). The opinion that 'free-range' and 'outdoor' systems are better for animal welfare than 'factory farming' has been challenged, with scientific evidence that these alternative systems have some important drawbacks and, in many aspects, fare worse than well-designed confined systems (Edwards, 2005; Fraser, 2008; Miele *et al.*, 2011). However, these criticisms miss the point: the relevant message is that the public considers industrial livestock systems unethical (New York Times, 2012). Free-range and outdoor systems are just the examples lay citizens can quote as alternatives; citizens are not expected to propose appropriate systems, but experts familiar with the systems are.

In this context, it is important to note that the question posed in the title of this chapter involves an ethical decision of what kind of science should be pursued. As Bernard Rollin (2006) discussed in detail, scientists resist the notion that science is a value-laden activity. However, one important ethical issue is what questions one attempts to answer. The values underpinning the kind of science that has been supported by different funding agencies and carried out by livestock scientists and farm animal welfare scientists are sometimes stated explicitly, but are more often 'between the lines' of published articles and books, and calls for research funding. Until recently, the main demand most animal welfare scientists have attempted to answer (implicitly if not explicitly) is how to produce abundant cheap food with acceptable animal welfare standards. The seemingly urgent need to achieve these goals and at the same time contribute to reducing the impacts of livestock production on the environment is a relatively new phenomenon in the history of livestock production science. Before embarking on more welfare science to help produce abundant cheap food in further intensified systems (i.e. 'sustainable intensification'), we should examine first whether this option accommodates realistic improvements in farm animal welfare, and second, whether it is the best or at least a good option to face the current and future challenges of agriculture.

5.4 Welfare Dilemmas

Remarkable advances in the understanding of animal sentience, and in the development and validation of scientific methods to assess farm animal welfare, have been achieved in the past 50 years (see review by Lawrence, 2008). Possibly the most notable improvement in farm animal welfare has come from changes in housing systems that prevented animals from meeting the five freedoms proposed by the Brambell Committee: to be able to stand up, lie down, turn around, groom themselves and stretch their limbs (Brambell, 1965). Housing that prevents such basic freedoms has been phased out in parts of the world, in some cases by laws or guidelines and in others via welfare assurance programmes.

However, factors other than housing can also affect the welfare of livestock reared in industrial systems; these include genetic selection for high productivity, restricted feeding, overstocking and painful husbandry practices. Many of the proposed methods for addressing these welfare issues within industrial production systems have involved trade-offs, and in some cases, solving one welfare problem can cause other welfare problems of equal or greater magnitude. These have been referred to in the literature as 'welfare dilemmas' (Mench, 2002; Lawrence, 2008; Dawkins and Layton, 2012).

Genetic selection has been one of the main drivers of increased livestock productivity. These increases in production efficiency can have economic benefits, but may also impose a toll on animals. Indeed, some have argued that selection for high production underlies many of the welfare problems of industrial livestock production (Hötzel and Machado Filho, 2004; Sandøe, 2008). Intensive selection for high production has resulted in a high incidence of debilitating metabolic disorders in livestock animals (Rauw *et al.*, 1998). For example, a myriad of metabolic diseases associated with rapid growth, increased metabolism or egg production affect the cardiovascular and musculoskeletal systems in poultry (Julian, 2005). Similarly, high-producing dairy cows are prone to develop mastitis, lameness, milk fever, retained placenta, ketosis, metritis and cystic ovaries (Gröhn *et al.*, 2003; Dobson *et al.*, 2007; Oltenacu and Broom, 2010). Selection for increased production in pigs has also imposed morphological and physiological imbalances that have resulted in leg disorders and reduced the animal's ability to cope with environmental challenges (Rauw *et al.*, 1998; Prunier *et al.*, 2010). Also in pigs, increasing average litter sizes through genetic selection has resulted in greater piglet mortality (Weber *et al.*, 2007; Rutherford *et al.*, 2013). There is a general consensus that such effects can reduce animal welfare due to hunger, pain or discomfort, and in some cases, impair reproduction and longevity (Rauw *et al.*, 1998; Dobson *et al.*, 2007; Decuypere *et al.*, 2010; Prunier *et al.*, 2010; de Jong and Guémené, 2011). However, initiatives and attempts to rethink productivity-oriented selection goals to include traits that could favour animal welfare in selection programmes have been limited (D'Eath *et al.*, 2010; Oltenacu and Broom, 2010; de Jong and Guémené, 2011).

Genetic selection for high productivity is often associated with high feed intake, and this can lead to high levels of hunger during certain stages of production. For example, both pregnant sows and broiler breeders (the parent stock of broiler chickens) are no longer at a life stage where high rates of weight gain are desirable, so these animals are typically fed only approximately 50% of voluntary intake (Mench, 2002; D'Eath *et al.*, 2009; Decuypere *et al.*, 2010). If given *ad libitum* access to the same concentrated diets typically provided, these animals would face increased risks of disease and reproductive failure; the dilemma is to choose between hunger and diseases related to obesity (Mench, 2002; Renema and Robinson, 2004; Dawkins and Layton, 2012; see also Sandøe *et al.*, Chapter 3, this volume).

Another dilemma revolves around two major, interrelated pig welfare problems: piglet death by crushing and the restriction of movement by sows confined in farrowing crates (Fig. 5.2). Farrowing crates were developed with the goal of reducing piglet crushing (Edwards, 2002), but these also prevent sows from engaging in most normal behaviours, including turning around and constructing a nest (Johnson and Marchant-Forde, 2009). Keeping sows in loose housing solves this problem, but often increases piglet mortality when piglets are crushed underneath the sow as she lies down in this more open

Fig. 5.2. Contrasting views of 'factory' and outdoor farming: (a) pregnant sow building a nest; (b) sow and litter in a farrowing crate; (c) sow and piglets interacting

(continued)

Fig. 5.2. (*Continued*) in an outdoor environment; and (d) piglet biting penmate. (Source of photos: Laboratório de Etologia Aplicada, UFSC, Brazil, with permission.)

enclosure (Johnson and Marchant-Forde, 2009; Vanheukelom *et al.*, 2012). Piglets that spend more time in contact with the sow are more likely to be crushed (Weary *et al.*, 1996), but changes in loose housing systems designed to encourage piglets to spend more time away from the sow (e.g. by offering a warm, soft 'creep' area attractive to the piglets) have, to date, met with little success (Vasdal *et al.*, 2009). This problem is further complicated by selection for larger litters, which has the effect of increasing the proportion of lighter piglets (Quiniou *et al.*, 2002; Quesnel *et al.*, 2008), which are more prone to be crushed by the sow (Pedersen *et al.*, 2011).

A final example of a welfare dilemma is the widespread practice of tail docking in piglets, with the aim of reducing the risk of tail biting (see Edwards and Bennett, Chapter 2, this volume). Tail biting is common in confined systems and can cause painful injuries and death in grower pigs (Schrøder-Petersen and Simonsen, 2001; Sinisalo *et al.*, 2012). Tail docking also causes pain (Marchant-Forde *et al.*, 2009), and methods of pain mitigation are not commonly used on-farm. In addition to the immediate pain associated with the procedure, neuromas can form at the site of the amputation, potentially causing chronic pain (Sutherland and Tucker, 2011). Tail docking would not be necessary if the underlying causes of tail biting were addressed. Factors associated with outbreaks of tail biting include large group sizes, high stocking densities, barren environments and genotype, although the relative contribution of each factor is poorly understood (Taylor *et al.*, 2010). Although European legislation (EC, 2001) mandates the use of relevant enrichment in housing and restricts the use of tail docking to exceptional cases, tail docking is still a common practice in many European countries (Compassion in World Farming, 2008), indicating the difficulty of controlling this abnormal behaviour in intensive systems. Sutherland and Tucker (2011, p. 189) conclude that tail 'docking tends to reduce tail biting, and until this behaviour is further understood and preventative measures adopted, it seems likely that amputation of the tail will continue to be used as a management practice'.

These are just a few of the many examples described in the literature of welfare problems that are difficult to solve at the farm level. It is important, though, to distinguish between the welfare dilemmas described above (where one welfare harm is traded against another) versus conflicts between welfare goals and other values (such as economics). An example of the latter dilemma is provided by Renema and Robinson (2004, p. 518) in reference to the 'broiler breeder paradox': 'realistically, genetic progress in broiler growth efficiency will not slow until economics warrant slowing the push to improved efficiency and yield'. Furthermore, value conflicts involving animal welfare will not only involve economic factors; the demand to produce more food in environmentally sustainable ways can value production efficiency over other goals, and as outlined below, the most efficient production systems are not always those that best promote welfare (Dawkins and Layton, 2012).

5.5 Will Environmental Issues Overshadow Animal Welfare?

Projected human population growth and income changes are expected to increase the demand for food, including the demand for meat products (Schneider *et al.*, 2011; Tilman *et al.*, 2011) and grains for animal feed (Keyzer *et al.*, 2005). Consequently, an intense debate surrounds the future of global food production, environmental sustainability and food security. One of the first and most thorough analyses of the negative implications of livestock production on the environment concludes:

> The livestock sector has such deep and wide-ranging environmental impacts that it should rank as one of the leading focuses for environmental policy ... Indeed, as societies develop, it is likely that environmental considerations, along with human health issues, will become the dominant policy considerations for the sector.
>
> (Steinfeld *et al.*, 2006, p. xxiv)

The harmful effects of livestock production, directly or through the crops required to feed animals, are land use change, excess water depletion and pollution, nutrient excretion, use of fossil energy, loss of biodiversity and emission of greenhouse gases (Steinfeld *et al.*, 2006). Possible scenarios discussed to mitigate these impacts involve reducing wastes and closing the gaps between current and potential yields (Godfray *et al.*, 2010; Foley *et al.*, 2011), shifting diets to a greater proportion of white meats and lower per capita meat consumption (Steinfeld *et al.*, 2006; Garnett, 2009; Wirsenius *et al.*, 2010; Henning, 2011; Berners-Lee *et al.*, 2012; see Appleby, Chapter 6, this volume) and 'sustainable intensification' of production. The latter may be summarized as the adoption of a series of technical, political and technology transfer strategies to increase productivity while minimizing negative environmental impacts (Steinfeld *et al.*, 2006; Royal Society, 2009; Foley *et al.*, 2011; Tilman *et al.*, 2011).

Reconciling the trade-offs between productivity and animal welfare under the approach of sustainable intensification may prove more difficult than under current production systems (Poletto and Hötzel, 2012). This is because sustainable intensification of livestock production is typically thought of as increased use of highly productive genotypes and large-scale confined production of monogastrics (Steinfeld *et al.*, 2006), likely to exacerbate the welfare issues already discussed above. Among the reasons provided for a shift to monogastrics (Gill *et al.*, 2010; Godfray *et al.*, 2010; Steinfeld and Gerber, 2010) is their higher feed conversion efficiency compared with most other livestock species, the associated lower economic and environmental costs of production and the acceptance of these meats, especially in regions of the world where demand for meat is expected to increase the most.

The title of this section highlights that the influential FAO report presented by Steinfeld *et al.* (2006) – *Livestock's Long Shadow* – as well as many recent analyses of strategies to overcome the challenge of producing more food with less environmental impact (Tilman *et al.*, 2002; Royal Society, 2009; Godfray *et al.*, 2010), have paid little or no attention to animal welfare. For instance, the Royal Society (2009, p. 39) stated that 'the sustainable intensification of agriculture requires a new understanding of these impacts so that interventions can be targeted to minimize adverse effects on the environment', omitting any mention of the potential impacts on animals. The dissociation between the environmental and farm animal welfare discussions occurs despite the growing evidence of the relevance of this issue to the public – or at least to the part of the public active enough to mobilize private and public institutions to regulate and effect changes in livestock production methods (Fraser, 2006; Centner, 2010). Some authors have argued that the potential uncertainty of the impacts of greenhouse gas mitigation options on animal welfare may delay decisions to act on environmental issues (de Boer *et al.*, 2011).

Policies that help to reduce the negative environmental impacts of agriculture, which are high on the international agenda, may conflict with welfare goals. While discussing the inclusion of welfare traits in breeding programmes of broiler chickens, Dawkins and Layton (2012) stated that we might not be able to 'have it all' (cheap food, environmental sustainability and animal welfare). The risk remains that policy makers will allow environmental issues to trump animal welfare concerns (Lawrence, 2008).

5.6 Attempts to Use Regulation to Promote the Adoption of Production Methods that Help Improve Farm Animal Welfare

The globalization of the food trade has increased market competition, amplifying the conflicts between welfare goals and economics that limit or prevent the adoption of practices that could improve farm animal welfare. Harmonizing production methods worldwide may avoid products from less regulated regions competing with those with stricter animal welfare regulation, minimizing the problems caused by price competition. This issue is becoming increasingly relevant, as the growth in livestock production and demand for meat will occur mostly in developing countries that already produce more than the half of the total world production (Steinfeld *et al.*, 2006; FAO, 2011a); these countries are emerging as important global food suppliers but generally do not have well-developed animal welfare regulations (Poletto and Hötzel, 2012). In recent years, there has been growing interest from countries with mature animal welfare regulatory frameworks to create and implement international regulation of farm animal welfare. This integration has taken place through organizations such as the World Organization for Animal Health (OIE) and the FAO, and corporations' quality assurance programmes (Fraser, 2006). However, as will be argued below, these mechanisms typically depend on the current economic environment of livestock production.

Some quality assurance programmes involve alternative farming methods like organic or free range, which may serve local and international markets through retailing chains. Although the demand for these products is growing rapidly throughout the world, it seems likely that conventionally produced products will continue to dominate the market for some time. These alternative systems compete with conventional systems that operate with economies of scale and low profit margins, and research suggests that few people are typically willing to pay premiums for animal welfare (Krystallis *et al.*, 2009; Centner, 2010; Uzea *et al.*, 2011; Lusk and Norwood, 2012).

How developing countries will respond to international initiatives to harmonize animal welfare standards is not yet known. One likely outcome is that exporting countries may respond with fitted-to-export product differentiation programmes, specifically organized to meet the demands of importing countries (Poletto and Hötzel, 2012). The industry in these countries is already prepared to deliver products that differ from those sold in the internal market, in order to meet specific demands from importing consumers. These include chickens slaughtered at a specific weight, pasture-based beef, cattle slaughtered in religious rituals, meat products free of antibiotics or certified as free of foot-and-mouth disease or free of the feed additive, ractopamine. Although this is a reasonable marketing strategy, it does little to help the animals destined for markets that do not demand welfare standards.

Finally, because these regulatory systems are tied to trade and competition, they may result in a reduction in the number of producers throughout the world, an outcome that contrasts with the known social relevance of livestock production in many regions of the world (FAO, 2009). This phenomenon, which occurred with the industrialization of livestock production in the last century (Tilman *et al.*, 2002; Fraser, 2008; Machado Filho *et al.*, 2010), may be furthered as an unintended result of welfare regulation, as some farmers are unable to invest the capital necessary to comply with the new standards (EC, 2006; Mench *et al.*, 2011).

5.7 To Improve Farm Animal Welfare, We Need Broad Change

Changes in current livestock production systems seem inevitable, as the required reductions in the environmental impact of livestock cannot be met by keeping 'business as usual' (Steinfeld *et al.*, 2006). Keeping animal welfare in the agenda of 'sustainable intensification' appears to be a significant challenge (Garnett *et al.*, 2013). Animal welfare is an important ethical social concern and, as such, needs to be integrated to the concept of sustainable agriculture, rather than made to 'compete' with environmental goals. A brighter future for farm animal welfare may be envisaged if the changes address an important

Fig. 5.3. Dairy cows grazing in a rotational system in a subtropical region of southern Brazil. When well managed, a pasture-based system can improve the welfare of cows and reduce environmental impacts compared with intensive grain-based systems. Additionally, lower dependence on external inputs may improve profit margins and shield farmers from price volatility. (Source of photo: Laboratório de Etologia Aplicada, UFSC, Brazil, with permission.)

approach to agriculture sustainability: the integration of animal, plant and soil systems (Fig. 5.3). Some ways to achieve this are by promoting mixed crop–livestock systems (Herrero *et al.*, 2009), moving from 'animal production' to 'animal husbandry' (Gjerris *et al.*, 2011) and pushing for changes aimed at 'a recoupling of crop and livestock systems ... through pricing and other policy mechanisms that reflect social costs of resource use and ecological abuse' (Naylor *et al.*, 2005, p. 1622). Or, as expressed by Janzen (2011, p. 791), linking livestock and land in 'a system based on loops, not lines'.

Achieving these goals calls for broad changes in practices, governance and society's values (Gjerris *et al.*, 2011; Gliessman, 2011; Janzen, 2011; Rosset, 2011). However, these barriers should not be understood as arguments to dismiss these ideas as unrealistic. A brief review of the recent literature reveals that the discussion of sustainable intensification is not so much about what new knowledge is needed to make intensive systems more efficient (i.e. sustainable); it is rather about what is required to implement this knowledge: strong political commitment, the ability to change society's values, to transfer 'appropriate' technologies to developing nations and international agreements to reform global trade policies (Tilman *et al.*, 2002, 2011; Steinfeld *et al.*, 2006; Foley *et al.*, 2011).

5.8 Conclusions

- The urgent need to reverse or mitigate the negative impacts of livestock production on the environment may increase pressure for further intensification of production systems, potentially harming animal welfare.
- Changes to production systems must consider animal welfare, human health and nutrition, the environment and food security.
- Citizen preferences for free-range or outdoor systems reflect moral outrage over the use of intensive confined methods. Instead of challenging this assessment, scientists should develop better alternatives that take these concerns into account.
- Ideally, solutions will integrate animal, plant and soil systems, potentially requiring broad changes in the production, trade and consumption of food. This does not imply a simplistic return to the past (see criticism of the 'agrarian romantic view' by Fraser, 2008). On the contrary, a change in food production must be underpinned by innovation, creativity and social ethics.
- Intensification without addressing society's concerns regarding the welfare of animals is unlikely to be sustainable.

Questions for discussion

1. Social concerns regarding farm animal welfare vary in different parts of the world. When products are sold in or exported to your country, you may, as a consumer, affect farming practices through your choice of products. Do consumers have a role to play here?
2. Should consumers eat less meat to accommodate more extensive production methods?
3. Livestock, especially ruminants, are essential for the livelihoods of many of the poorer people living in developing countries. How should these food security concerns be balanced with animal welfare and environmental concerns?
4. Is purchasing animal food products from alternative production systems a satisfactory response to moral concerns regarding farm animal welfare? Does it also help to prevent the suffering of animals reared in conventional systems?

Acknowledgements

I acknowledge the critical and constructive comments to previous versions of this chapter made by Michael Appleby, João Henrique Cardoso Costa, Alfredo Celso Fantini, Peter Sandøe, Ricardo Rüther, Daniel Weary and the authors of other chapters of this book who participated in the workshop held in Vienna in August 2012. I also acknowledge the financial support of the Universidade Federal de Santa Catarina to attend the Workshop in Vienna and of CNPq, Brasília, Brazil, Proc. 308919/2009-2.

References

Berners-Lee, M., Hoolohan, C., Cammack, H. and Hewitt, C.N. (2012) The relative greenhouse gas impacts of realistic dietary choices. *Energy Policy* 43, 184–190.

Brambell, F.W.R. (1965) *Report of the technical committee to enquire into the welfare of animals kept under intensive livestock husbandry systems*. HMSO, London.

Centner, T.J. (2010) Limitations on the confinement of food animals in the United States. *Journal of Agricultural and Environmental Ethics* 23, 469–486.

Compassion in World Farming (2008) *The State of Europe's Pigs: An Exposé. An Investigative Report by Compassion in World Farming*. Compassion in World Farming, UK (http://www.farmsnotfactories.org/pdfs/CIWF-state_of_europes_pigs.pdf, accessed 8 August 2013).

Dawkins, M.S. and Layton, R. (2012) Breeding for better welfare: genetic goals for broiler chickens and their parents. *Animal Welfare* 21, 147–155.

D'Eath, R.B., Tolkamp, B.J., Kyriazakis, I. and Lawrence, A.B. (2009) 'Freedom from hunger' and preventing obesity: the animal welfare implications of reducing food quantity or quality. *Animal Behaviour* 77, 275–288.

D'Eath, R.B., Conington, J., Lawrence, A.B., Olsson, I.A.S. and Sandøe, P. (2010) Breeding for behavioural change in farm animals: practical, economic and ethical considerations. *Animal Welfare* 19, S17–S27.

de Boer, I.J.M., Cederberg, C., Eady, S., Gollnow, S., Kristensen, T., Macleod, M., *et al.* (2011) Greenhouse gas mitigation in animal production: towards an integrated life cycle sustainability assessment. *Current Opinion in Environmental Sustainability* 3, 423–431.

Decuypere, E., Bruggeman, V., Everaert, N., Li, Y., Boonen, R., De Tavernier, J., *et al.* (2010) The broiler breeder paradox: ethical, genetic and physiological perspectives, and suggestions for solutions. *British Poultry Science* 51, 569–579.

de Jong, I.C. and Guémené, D. (2011) Major welfare issues in broiler breeders. *World's Poultry Science Journal* 67, 73–82.

Dobson, H., Smith, R., Royal, M., Knight, C. and Sheldon, I. (2007) The high-producing dairy cow and its reproductive performance. *Reproduction in Domestic Animals* 42, 17–23.

EC (European Commission) (2001) Commission Directive 2001/93/EC of 9 November 2001 amending Directive 91/630/EEC laying down minimum standards for the protection of pigs. *Official Journal of the European Communities* L316, 1–3 (http://eur-lex.europa.eu/LexUriServ/LexUriServ.do?uri=CELEX:32001L0093:en:NOT, accessed 1 November 2012).

EC (2006) Directorate General for 'Health and Consumers' of European Commission. Study on the socio-economic implications of different aspects of farming weaners and pigs kept for fattening (Aragrande, M. *et al.*) Final report (http://ec.europa.eu/food/animal/welfare/farm/farming_pigs_finalreport_en.pdf, accessed 8 April 2013).

EC (2007) Attitudes of EU citizens towards animal welfare. Special Eurobarometer 270/Wave 66.1 – TNS opinion and social (ec.europa.eu/food/animal/welfare/survey/index_en.htm, accessed November 2012).

Edwards, S.A. (2002) Perinatal mortality in the pig: environmental or physiological solutions? *Livestock Production Science* 78, 3–12.

Edwards, S.A. (2005) Product quality attributes associated with outdoor pig production. *Livestock Production Science* 94, 5–14.

FAO (Food and Agriculture Organization) (2009) The state of food and agriculture – livestock in the balance (www.fao.org/docrep/012/i0680e/i0680e.pdf, accessed 21 May 2012).

FAO (2010) The state of food insecurity in the world (www.fao.org/docrep/013/i1683e/i1683e.pdf, accessed 21 May 2012).

FAO (2011a) How to feed the world in 2050 (www.fao.org/fileadmin/templates/wsfs/docs/expert_paper/How_to_Feed_the_World_in_2050.pdf, accessed 21 May 2012).

FAO (2011b) FAOSTAT (http://faostat.fao.org/site/339/default.aspx, accessed 27 December 2012).

Foley, J.A., Ramankutty, N., Brauman, K.A., Cassidy, E.S., Gerber, J.S. Johnston, M., *et al.* (2011) Solutions for a cultivated planet. *Nature* 478, 337–342.

Fraser, D. (2006) Animal welfare assurance programs in food production: a framework for assessing the options. *Animal Welfare* 15, 93–104.

Fraser, D. (2008) Animal welfare and the intensification of animal production. In: Thompson, P.B. (ed.) *The Ethics of Intensification*. Springer Verlag, Heidelberg, Germany, pp. 167–189.

Garnett, T. (2009) Livestock-related greenhouse gas emissions: impacts and options for policy makers. *Environmental Science and Policy* 12, 491–504.

Garnett, T., Appleby, M.C., Balmford, A., Bateman, I.J., Benton, T.G., Bloomer, P., Burlingame, B., Dawkins, M., Dolan, L., Fraser, D., Herrero, M., Hoffmann, I., Smith, P., Thornton, P.K., Toulmin, C., Vermeulen, S.J. Godfray, H.C.J. (2013) Sustainable intensification in agriculture: premises and policies. *Science* 341, 33-34.

Gill, M., Smith, P. and Wilkinson, J.M. (2010) Mitigating climate change: the role of domestic livestock. *Animal* 4, 323–333.

Gjerris, M., Gamborg, C., Röcklinsberg, C. and Anthony, A. (2011) The price of responsibility: ethics of animal husbandry in a time of climate change. *Journal of Agricultural and Environmental Ethics* 24, 331–350.

Gliessman, S. (2011) Agroecology and food system change. *Journal of Sustainable Agriculture* 35, 347–349.

Godfray, H.C.J., Beddington, J.R., Crute, I.R., Haddad, L., Lawrence, D., Muir, J.F., *et al.* (2010) Food security: the challenge of feeding 9 billion people. *Science* 327, 812–818.

Gröhn, Y.T., Rajala-Schultz, P.J., Allore, H.G., Delorenzo, M.A., Hertl, J.A. and Galligan, D.T. (2003) Optimizing replacement of dairy cows: modeling the effects of diseases. *Preventive Veterinary Medicine* 61, 27–43.

Harrison, R. (1964) *Animal Machines*. Vincent Stuart, London.

Henning, B. (2011) Standing in livestock's 'long shadow': the ethics of eating meat on a small planet. *Ethics and the Environment* 16, 63–93.

Herrero, M., Thornton, P.K., Gerber, P. and Reid, R.S. (2009) Livestock, livelihoods and the environment: understanding the trade-offs. *Current Opinion in Environmental Sustainability* 1, 111–120.

Hötzel, M.J. and Machado Filho, L.C.P. (2004) Bem-estar animal na agricultura do século XXI. *Revista de Etologia* 6, 3–16.

Janzen, H.H. (2011) What place for livestock on a re-greening earth. *Animal Feed Science and Technology* 166–167, 783–796.

Johnson, A.K. and Marchant-Forde, J.N. (2009) Welfare of pigs in the farrowing environment. In: Marchant-Forde, J.N. (ed.) *The Welfare of Pigs*. Springer, Dordrecht, the Netherlands, pp. 141–188.

Julian, R.J. (2005) Production and growth related disorders and other metabolic diseases of poultry: a review. *Veterinary Journal* 169, 350–369.

Keyzer, M.A., Merbis, M.D., Pavel, I.F.P.W. and van Wesenbeeck, C.F.A. (2005) Diet shifts towards meat and the effects on cereal use: can we feed the animals in 2030? *Ecological Economics* 55, 187–202.

Krystallis, A., Barcellos, M.D. de, Kugler, J.O., Verbeke, W. and Grunert, K.G. (2009) Attitudes of European citizens towards pig production systems. *Livestock Science* 126, 46–56.

Lawrence, A.B. (2008) Applied animal behaviour science: past, present and future. *Applied Animal Behaviour Science* 115, 1–24.

Lusk, J.L. and Norwood, F.B. (2011) Animal welfare economics. *Applied Economic Perspectives and Policy* 33(4), 463–483.

Lusk, J.L. and Norwood, F.B. (2012) Speciesism, altruism and the economics of animal welfare. *European Review of Agricultural Economics* 39, 189–212.

Machado Filho, L.C.P., Hötzel, M.J., Pinheiro Machado, L.C.P. and Ribas, C.E.D. (2010) Transição para uma pecuária agroecológica. In: Lana, R.P., Guimarães, G., Veloso, C.M., Machado, T.M.M., Souza, M.R.M., Mancio, A.B., de Lima, D.V. and da Silva, J.C.P.M. (eds) *II, Simpósio Brasileiro de Agropecuária Sustentável*. Arka Editora, Viçosa, Brazil, pp. 245–260.

Marchant-Forde, J.N., Lay, D.C. Jr, McMunn, K.A., Cheng, H.W., Pajor, E.A., Marchant-Forde, R.M. (2009) Postnatal piglet husbandry practices and well-being: the effects of alternative techniques delivered separately. *Journal of Animal Science* 87, 1479–1492.

Mench, J.A. (2002) Broiler breeders: feed restriction and welfare. *World's Poultry Science Journal* 58, 23–29.

Mench, J.A., Sumner, D.A. and Rosen-Molina, J.T. (2011) Sustainability of egg production in the United States – the policy and market context. *Poultry Science* 90, 229–240.

Miele, M. (2010) Report concerning consumer perceptions and attitudes towards farm animal welfare. Official Experts Report EAWP (task 1.3). Uppsala University, Uppsala, Sweden.

Miele, M., Veissier, I., Evans, A. and Botreau, R. (2011) Animal welfare: establishing a dialogue between science and society. *Animal Welfare* 20, 103–117.

Naald, B.V. and Cameron, T.A. (2011) Willingness to pay for other species' well-being. *Ecological Economics* 70, 1325–1335.

Naylor, R. (2011) Expanding the boundaries of agricultural development. *Food Security* 3, 233–251.

Naylor, R., Steinfeld, H., Falcon, W., Galloway, J., Smil, V., Bradford, E., *et al.* (2005) Losing the links between livestock and land. *Science* 310, 1621–1622.

New York Times (2012) The meat you eat (www.nytimes.com/2012/05/06/magazine/the-winner-of-our-contest-on-the-ethics-of-eating-meat.html?_r=2&src=rechp, accessed 21 May 2012).

Nonhebel, S.and Kastner, T. (2011) Changing demand for food, livestock feed and biofuels in the past and in the near future. *Livestock Science* 139, 3–10.

Oltenacu, P.A. and Broom, D.M. (2010) The impact of genetic selection for increased milk yield on the welfare of dairy cows. *Animal Welfare* 19, S39–S49.

Pedersen, L.J., Berg, P., Jørgensen, G. and Andersen, I.L. (2011) Neonatal piglet traits of importance for survival in crates and indoor pens. *Journal of Animal Science* 89, 1207–1218.

Poletto, R. and Hötzel, M.J. (2012) The five freedoms in the global animal agriculture market: challenges and achievements as opportunities. *Animal Frontiers* 2, 22–30.

Prunier, A., Heinonen, M. and Quesnel, H. (2010) High physiological demands in intensively raised pigs: impact on health and welfare. *Animal* 4, 886–898.

Quesnel, H., Brossard, L., Valancogne, A. and Quiniou, N. (2008) Influence of some sow characteristics on within-litter variation of piglet birth weight. *Animal* 2, 1842–1849.

Quiniou, N., Dagorn, J. and Gaudré, D. (2002) Variation of piglets' birth weight and consequences on subsequent performance. *Livestock Production Science* 78, 63–70.

Rauw, W.M., Kanis, E., Noordhuize-Stassen, E.N. and Grommers, F.J. (1998) Undesirable side effects of selection for high production efficiency in farm animals: a review. *Livestock Production Science* 56, 15–33.

Renema, R.A. and Robinson, F.E. (2004) Defining normal: comparison of feed restriction and full feeding of female broiler breeders. *World's Poultry Science Journal* 60, 508–522.

Rollin, B.E. (2006) *Science and Ethics.* Cambridge University Press, New York.

Rosset, P. (2011) Preventing hunger: change economic policy. *Nature* 479, 472–473.

Royal Society (2009) *Reaping the Benefits: Science and Sustainable Intensification of Global Agriculture.* Royal Society, London.

Rutherford, K.M.D., Baxter, E.M., D'Eath, R.B., Turner, S.P., Arnott, G., Roehe, R., Ask, B., Sandøe, P., Moustsen, V.A., Thorup, F., Edwards, S.A., Berg, P. and Lawrence, A.B. (2013) The welfare implications of large litter size in the domestic pig I: biological factors. *Animal Welfare* 22, 199-218.

Sandøe, P. (2008) Re-thinking the ethics of intensification for animal agriculture: comments on David Fraser, animal welfare and the intensification of animal production. In: Thompson, P.B. (ed.) *The Ethics of Intensification.* Springer Verlag, Heidelberg, Germany, pp. 191–198.

Schneider, U.A., Havlík, P., Schmid, E., Valin, H., Mosnier, A., Obersteiner, M., *et al.* (2011) Impacts of population growth, economic development, and technical change on global food production and consumption. *Agricultural Systems* 104, 204–215.
Schrøder-Petersen, D.L. and Simonsen, H.B. (2001) Tail biting in pigs. *Veterinary Journal* 162, 196–201.
Sinisalo, A., Niemi, J.K., Mari Heinonen, M. and Anna Valros, V. (2012) Tail biting and production performance in fattening pigs. *Livestock Science* 143, 220–225.
Steinfeld, H.H. and Gerber, P. (2010) Livestock production and the global environment: consume less or produce better? *Proceedings of the National Academy of Sciences*107, 18237–18238.
Steinfeld, H., Gerber, P., Wassenaar, P., Castle, V., Rosales, M. and De Haan, D. (2006) *Livestock's Long Shadow: Environmental Issues and Options*. FAO, Rome.
Sutherland, M.A. and Tucker, C.B. (2011) The long and short of it: a review of tail docking in farm animals. *Applied Animal Behaviour Science* 135, 179–191.
Taylor, N.R., Main, D.C.J., Mendl, M. and Edwards, S.A. (2010) Tail-biting: a new perspective. *Veterinary Journal* 186, 137–147.
Tilman, D., Cassman, K.G., Matson, P.A., Naylor, R. and Polasky, S. (2002) Agricultural sustainability and intensive production practices. *Nature* 418, 671–677.
Tilman, D., Balzer, C., Hill, J. and Befort, B.L. (2011) Global food demand and the sustainable intensification of agriculture. *Proceedings of the National Academy of Sciences* 50, 20260–20264.
Uzea, A.D., Hobbs, J.E. and Zhang, J. (2011) Activists and animal welfare: quality verifications in the Canadian pork sector. *Journal of Agricultural Economics* 62, 281–304.
Vanheukelom, V., Driessen, B. and Geers, R. (2012) The effects of environmental enrichment on the behaviour of suckling piglets and lactating sows: a review. *Livestock Science* 143, 116–131.
Vasdal, G., Andersen, I.L. and Pedersen, L.J. (2009) Piglet use of creep area – effects of breeding value and farrowing environment. *Applied Animal Behaviour Science* 120, 62–67.
Weary, D.M., Pajor, E.A., Thompson, B.K. and Fraser, D. (1996) Risky behavior by piglets: a trade off between feeding and risk of mortality by maternal crushing? *Animal Behaviour* 51, 619–624.
Weber, R., Keli, N., Fehr, M. and Horat, R. (2007) Piglet mortality on farms using farrowing systems with or without crates. *Animal Welfare* 16, 277–279.
Wirsenius, S., Azar, C. and Berndes, G. (2010) How much land is needed for global food production under scenarios of dietary changes and livestock productivity increases in 2030? *Agricultural Systems* 103, 621–638.

Whom Should We Eat? Why Veal Can Be Better for Welfare than Chicken

6

Michael C. Appleby*
World Society for the Protection of Animals, UK

6.1 Abstract

Some people express their concerns about the suffering and death of farm animals by becoming vegetarian or vegan, but others think that eating food from sentient animals is justifiable. Such food should come from animals with the best possible welfare, and this is affected by people's food choices. Welfare can be improved by reducing animal numbers and looking after those animals more carefully, with increased attention to individuals. Some of the most important factors that affect this are body size, longevity, group size, the amount of animal products that people eat and the price the farmer receives. As an example, I argue that relatively long-lived, high-value cattle such as rose veal calves are more likely to have a good life than short-lived, low-value chickens. I propose that to safeguard welfare it is best to choose 'food with a name', from a system or farm in which the animals are actually, or might potentially be, given a name. Other names are also important, such as the name of the farmer or farm that produces the food and the name of the brand under which it is sold.

6.2 Introduction

An important aspect of addressing animal welfare is to consider animals as individuals: what are the factors that increase or decrease the quality of a

*E-mail: michaelappleby@wspa-international.org

particular animal's life? Some commentators think that the language we use about animals is a vital part of this. They suggest that using words such as 'it' or 'that' to refer to animals rather than 'he', 'she' or 'who' makes people more likely to treat them as objects rather than sentient beings (Hearne, 1982). Yet a question like 'Whom should we eat?' causes discomfort. It confronts the dilemma of caring for the welfare of individual animals while sanctioning their killing for food. It may cause more than discomfort. It may invite opprobrium from people to whom such killing is unacceptable and who therefore argue that animals should not be kept or killed for food at all. The question is used here to initiate discussion on the effects of our choices on animal welfare: our choice of whether to eat animal products and, if we do, our choice of the animals from whom those products should come.

6.3 Vegetarians and Vegans

The welfare of farm animals has been a subject of concern to many people in many countries, at least since the 1960s, when Harrison (1964) publicized the suffering caused by some farming methods, particularly new intensive husbandry methods (although see Weary, Chapter 11, this volume, for a discussion of what constitutes suffering). In recent years, there have been some improvements in farm animal welfare, for example through legislation in many developed countries, but there is still cause for concern (FAWC, 2009).

For some people, the obvious response to issues of either farm animal suffering or killing is vegetarianism. This reason for vegetarianism is not always distinct from other reasons such as health (Hill, 1996), but in this context a person's primary motivation is generally stated as to reduce their personal contribution to suffering and killing.

All the choices discussed in this chapter probably involve a mixture of personal and public motivations. Someone who is concerned for the animals that would have supplied them with food is also likely to want an end to the suffering and killing of all farm animals. However, for the rest of this chapter emphasis will be placed on the direct effect of choices rather than their indirect use for publicity or other public purposes.

Abstaining from the eating of meat does indeed reduce the personal contribution of vegetarians to suffering and killing. However, it does little directly to improve the treatment of those animals that continue to be farmed. An increase in the number of vegetarians does have an influence on the livestock industry and is one component in the changes in attitude, social pressures and development of legislation that have led to changes in how animals are treated. But vegetarians are still too few, at least in most developed countries, for this to be a large influence.

The animals that continue to be farmed include those providing the milk and eggs that most vegetarians also eat. And commercial milk and egg production involves not just the farming and killing (and any associated suffering) of dairy cows and laying hens, but often the killing of male calves and chicks that are judged to be useless for production. There are ethical differences between eating eggs and bacon (to quote the old phrase, 'the hen is involved but the pig is committed'), but they are not as great as they first seem. There is, therefore, some inconsistency between the intention of vegetarians to prevent the suffering and killing of farm animals and the actual effects of their diet.

Veganism – abstaining from all animal products – goes further to avoid inconsistency. Some inconsistency persists, though, including the fact that some animals are killed, controlled or displaced for the purposes of crop production. It has even been claimed that more animals are killed to sustain a vegan diet than an omnivorous one (Davis, 2003), although this is disputed (Matheny, 2003). In any case, inconsistencies do not invalidate an ethical approach. Hill defends a vegetarian who swats a fly or wears leather shoes from the charge of inconsistency or hypocrisy on the grounds that:

> Complete consistency is not always possible in the real world. What is important is that she is making a real effort to live by her beliefs. In short, she may not always be consistent, but is the only other alternative to absolute consistency to make no attempt at all? Is it not better to make even modest strides in the direction to which one aspires, even if one cannot always reach one's aspiration, than to make no attempt at all?
>
> (Hill, 1996, p. 168)

So, even though it is impossible to avoid all negative impacts on animals, a vegan is making a considerable effort to minimize his or her own impact.

An option potentially open to vegans (and others) is artificially cultured meat. There is increasing interest in producing such meat, to reduce both the welfare and the environmental problems of livestock (Olsson, 2008), but it is not clear whether or when it might be practical.

An argument sometimes made against veganism is that if everyone was vegan this would mean that farm animals would be almost completely eliminated, reduced to a few kept as pets, in zoos or in managed wild areas. The argument suggests that this would be both impractical and undesirable, entailing the loss not just of the animals but also of our symbiotic relationship with them that has been established for thousands of years (although this is not an argument about the welfare of individual animals). However, while vegetarians are few in number, vegans are fewer. Choosing to be vegan does not require someone to justify a hypothetical world where everyone is vegan. Such a world is not a foreseeable option. Veganism is, again, principally a personal choice. Nevertheless, the observation that a world without farm animals would be impractical and/or undesirable is a reminder that while the dietary choices of individual people are personal, those of a population have widespread consequences and are part of the complex picture of the population's interactions with its environment. Livestock and their manure are an important element in mixed farming, including organic agriculture. Livestock can feed on marginal land and on plants that cannot be used for human food production. In many countries, livestock are also important for labour, fuel, clothing, social status and security.

So, while vegetarians and vegans reduce their personal contribution to the suffering and killing of livestock compared to other people, animals will continue to be farmed, and their welfare remains an issue.

6.4 Conscientious Omnivores and Vegetarians

For people who eat animal products (including vegetarians) but who are concerned about animal welfare, the next option to be considered is selectivity in the welfare provenance of those products. Singer and Mason (2006) have popularized the term 'conscientious omnivore' in this context, and it would also be possible to consider conscientious vegetarians (and indeed vegans, as the conscientiousness may also concern ecological, social and other effects of diet).

Someone who wants to choose products from animals whose welfare is better than average generally does so by selecting food labelled as organic, free range or specifically humane, or from a source (such as a local farm) that they believe provides this guarantee. Not all animals in such systems will necessarily have above-average welfare. Concerns are sometimes raised, for example, about organic dairy cows with mastitis left untreated because treatment would disqualify them from organic sales. However, there is no suggestion that the majority of organic animals have worse welfare than others. All animals in such programmes do have access to extra resources which (at least potentially) improve their welfare over those farmed conventionally, such as extra space allowing increased freedom of movement. It is probably fair to say that for the majority of these animals such advantages outweigh the disadvantages. So, these purchasers will ensure that the animals supplying them with food have, on average, had a better life than others. In supporting the farms and businesses supplying that food, they will also ensure that other animals in future have a better life than they otherwise would have.

How much better is their life? Well, all voluntary programmes intended to improve farm animal welfare have to decide how stringent their criteria should be and, in general, those with high standards are likely to attract fewer farmers and help fewer animals, while those with lower standards will achieve a larger market. Variation in the demands made by different groups is illustrated by Norwood and Lusk:

> Some of the issues promoted by HSUS [The Humane Society of the United States] are not radical at all, and if the votes on ballot initiatives in Florida, Arizona and California are any indication, its position on issues such as giving animals room to lie down, stand up, and fully extend their limbs are popular with the American public.

> Now consider the Animal Welfare Institute (AWI) ... In our opinion, the AWI standards are among the highest animal welfare standards that exist for commercial farms today.
>
> (Norwood and Lusk, 2011, p. 59)

However, while the HSUS is attempting to achieve improvements for all farm animals, relatively few farms work to AWI standards. All purchasers of 'higher welfare' animal products are making a choice on this continuum from small improvement for many animals to large improvement for few. It is worth noting, though, that they are usually doing so with incomplete information, because food labels will not help them fully on this. All such labels may be expected to say something like 'guaranteed humane'. They cannot be expected to say anything like 'guaranteed slightly more humane'.

This approach, of choosing products from animals receiving better than average treatment, is having a widespread impact. Niche markets have led the way to wider change. The growth in the purchase of free-range eggs, for example, even though such eggs were a minority of the total, was apparently taken by politicians in the European Union and many of its member countries to indicate sufficient support to proceed with legislative change on the housing of laying hens (Appleby, 2003).

Others have criticized this 'incremental' approach as not going far enough. Some authors conclude that even if animals' lives are improved, their killing is still not justified (Foer, 2009). Others argue that incremental improvements should not be accepted in a system that needs larger change (see Francione's contributions to Francione and Garner, 2010). This reasoning suggests that increasing the size of chicken cages (and even adding enrichment) may do more harm than good by helping to perpetuate the commercial keeping of chickens. Haynes suggests that actions such as increasing cage size are based on a misconception of animal welfare that plays ...

> a major role in supporting merely limited reform in the use of animals and seems to support the assumption that there are conditions under which animals may be raised and slaughtered for food that are ethically acceptable. Reformists do not need to make this assumption, but they tend to conceptualize animal welfare in such a way that death does not count as harmful to the interests of animals, nor prolonged life a benefit.
>
> (Haynes, 2011, p. 105)

Haynes is arguing that incremental changes to rearing conditions are based on a false premise and fail to address the most important issue, that rearing and slaughtering animals may itself be ethically unacceptable.

Still, for many people the lifelong welfare of animals is more important than the instant of death, particularly if the death can be humane, and any improvement in their welfare is better than none. Providing a good life and a gentle death are both important aspects of taking responsibility for animals in our care (Pollan, 2006).

So far, choices about animal products have been discussed as if the implications for all animals were the same. They are not.

6.5 Choice Between Invertebrates, Fish, Birds and Mammals

This chapter began by emphasizing the importance of considering animals as individuals, and in that sense the category of animals from which food products come should be irrelevant: what matters is the welfare of the animals and that it is improved as far as possible. This approach tends to exclude insentient animals. 'Sentience' is used here to mean 'capacity for suffering or pleasure'. There is no rigid dividing line between sentient and insentient species (Appleby, 1999), and there is debate about the sentience of some groups such as cephalopods and decapod crustaceans. There is probably general agreement, or at least a large majority opinion, that some invertebrate groups, such as insects, do not have the capacity for suffering that is comparable to the suffering of vertebrates. As such, there is little, if any, need to be concerned about the welfare of animals in these groups (although there may well be other ethical concerns about their treatment). Animals from some of these groups, like insects and shellfish, are quite widely used as food in at least some parts of the world, but further discussion here shall be restricted to sentient species, especially vertebrates.

There is increasing consensus and scientific evidence that all vertebrates and some invertebrates can be categorized as sentient species. Among vertebrates, there has been most debate about fish, but they are now usually included in the consensus (Braithwaite, 2010). There is certainly increasing belief that, in at least some treatments relevant to welfare, fish, cephalopods and decapods should be given the benefit of the doubt. Much commercial slaughter of farmed fish is now preceded by stunning (Branson, 2008), and the practice of killing crabs and lobsters by placing them in boiling water is being replaced by alternative methods in some countries (Roth and Øines, 2010).

Yet some people are more willing to eat fish than mammals or poultry, and worry less about their welfare. Is this because their welfare is satisfactory?

The welfare of farmed fish is increasingly considered (Branson, 2008). For example, there is growing understanding of the parameters of water quality that are important for welfare, and control of these parameters in practice. Such factors are included in welfare programmes such as the RSPCA Freedom Food standards for salmon (RSPCA, 2011). Yet it is still difficult to see the welfare of farmed fish, even on such programmes, as equivalent to those of the best cared for mammals and poultry. They still largely swim in circles in featureless tanks or nets holding large numbers. Such environments are seen as necessary for commercial production. It is true that there is little scientific evidence that welfare is affected by these conditions, but then there has been little research on this, including on what might be expected to be some of the most important features. Those features will vary between species, but to take an example, salmon – the most common species farmed in the UK – are carnivorous and migratory, yet almost nothing is known of the effects of preventing hunting and migration on their welfare. There are also many significant welfare problems for farmed fish, such as diseases that are difficult to control, and mortality is often disturbingly high. While the public is largely unaware of such problems, it seems unlikely that a general lack of concern for fish welfare is because of confidence that it is high.

Perhaps a more likely cause of such lack of concern is that people have less empathy with fish than with warm-blooded animals. Even when people accept that fish experience negative states such as pain, they still often describe them as somehow alien and respond less to descriptions of fish suffering than to equivalent descriptions of species more similar to humans. For example, billions of fish are killed annually by suffocation by removing them from water (Mood, 2010), yet people do not object to this as strongly as they would to equivalent treatment of birds or mammals, or try so hard to prevent it. Rod fishing is acceptable in most societies, despite using a method of capture – a hook in the mouth of an animal fighting to escape – that would be unacceptable for a bird or a mammal.

The same may be true of attitudes to birds compared with that to mammals. Some people think the welfare of poultry is less important than that of mammals, and again this may be due to lack of empathy. Chickens are sometimes described as 'reptilian', often dismissed as 'bird brained'. Yet the fact that fish and chickens are very different from us, with a lack of facial expressions and behavioural responses different from our own, is no reason to believe that they have a lower inherent capacity to suffer than other mammals or that they are suffering less than mammals in the same situation.

Indeed, as for fish, there are good reasons to be concerned about chicken welfare (Fig. 6.1), even in systems promoted specifically as welfare friendly, such as free range. In commercial free-range production, both broilers and layers are kept in very large flocks, with consequent limitations on the complexity of their environment, on their ability to interact with it and hence on their welfare. There is also very little done to identify and treat welfare problems in individual birds in such flocks.

By buying fish and poultry products, people are supporting the large-scale simple production systems in which these animals are kept, associated with their relatively small body size. I shall return to the issue of body size in relation to animal numbers after the next section.

6.6 Choice Between Other Categories of Animal

Only large taxonomic categories of animals have been considered so far but when choices within or independent of those categories are considered, there are obvious overlaps or contradictions with choices on other criteria. As before,

Fig. 6.1. Most chicken meat sold in most countries is from broiler chickens reared in very large flocks, stocked at high density in simple housing, with almost no possibility for attention to the welfare of individual birds. (Photograph from the United States Department of Agriculture. Used with permission.)

motivations for dietary choices are mixed, including animal welfare, health, ecology and the social effects of diet.

Thus, some people tend to choose white rather than red meat, because of a belief that it is healthier. This usually means choosing meat from pigs, chickens or fish (although there are additional health reasons for choosing fish) rather than from cattle or sheep. In developed countries, the former are mostly kept in intensive conditions, while sheep and some beef cattle are in extensive conditions (although there are also many cattle in feedlots), with implications for welfare. Also of note is that pigs and chickens are monogastrics, whereas cattle and sheep are ruminants. Pigs and chickens often eat grain that people could eat, an issue important both to environmental impact and to food security, while ruminants can use marginal land and plants unsuitable for human food production (although many cattle are also fed on grain). Thus, someone choosing white meat rather than red does so despite at least some arguments favouring the opposite choice, concerning both environmental impact and the fact that the animals producing white meat are kept intensively, with likely negative effects on their welfare.

Clearly, there are many other choices among animal products that affect the welfare of the animals from which those products come. These include choices between species, which often have very different rearing conditions with profound effects on welfare. For example, ducks and geese have different requirements for water from those of chickens, yet these are often not met in commercial management systems. Choices are also influential within species, between different categories of animals (for example, veal calves and other cattle) or between younger and older animals. However, while there is a huge number of factors that affect whether a particular category of animals or individual has good welfare in its life or prior to death, no one choosing an animal product is likely to consider all those factors separately. Rather, a conscientious omnivore or vegetarian concerned for the welfare of the animals supplying them with food is likely to use broad criteria for his or her choices, as already discussed, criteria such as humane or organic labelling, free range or caged, fish or mammal. Factors such as breed, specific husbandry practices and slaughter method are relevant to those broad criteria but will less often determine the actual choice, unless they have come to public attention, perhaps through a recent news story.

One factor that needs more attention than it has received hitherto, though, is animal numbers, to which I now turn.

6.7 Numbers of Animals

In 2010, approximately 70 billion terrestrial farm animals were killed for food, including dairy animals and poultry producing eggs for consumption (Table 6.1, FAO, undated), and of the order of 80 billion farmed fish (Mood and Brooke, 2012). Of the former figure, 66 billion were poultry and 3.4 billion were mammals.

Table 6.1. Farm animals (excluding fish) slaughtered for food in 2010, millions (FAO, undated).

Chickens	61,834
Ducks	2,708
Turkeys	632
Other poultry	809
Poultry	*65,983*
Pigs	1,375
Cattle	559
Buffaloes	82
Sheep	749
Goats	617
Camelids	8
Equids	8
Rabbits	1
Mammals	*3,399*
Total	*69,382*

If those numbers are turned back into consumer choices, people are eating the meat of nearly 20 birds for every mammal. Clearly, there are many factors of supply and demand that affect those numbers both regionally and globally, but one of the most important is body size. One cow supplies approximately 200 times as much meat as a chicken, so if people bought the same amount of chicken and beef, that would require 200 times as many chickens to be reared and killed as cows. In fact, the ratio is about 110 chickens to one cow.

The implications of this calculus for welfare are complex. Those who emphasize the suffering and killing of farm animals, consequent on the eating of meat, conclude that the negative consequences of rearing and slaughtering this huge number of poultry outweigh those for the much smaller number of mammals. Ingrid Newkirk famously and provocatively suggested that people who wanted to eat animal products should eat whale meat:

> You can get thousands and thousands of meals out of a whale, so you wouldn't be killing and torturing so many of them. Seriously, I think everybody needs to be more disciplined; nobody needs any meat. But from a perspective of how many animals suffer, it's probably better to kill and eat one whale than it is to eat fish, chickens, cows, lambs and eggs.
>
> (Newkirk, undated)

Newkirk's suggestion is ironic but, for the record, eating whales instead of farm animals is not sustainable. A blue whale would yield about 150 times as much meat as a cow (Anon., 2011) (so about 30,000 times as much as a chicken) but 559 million cows are slaughtered annually worldwide (Table 6.1), equivalent to 3.7 million blue whales. That is approximately 400 times the world population.

More to the point, the suffering and killing of farm animals are not the only parameters important to their welfare. There are also positive aspects and periods in their lives.

FAWC (2009) suggests that an animal may have either 'a life not worth living', 'a life worth living' or 'a good life'. How many farm animals have a life not worth living, such that they 'would be literally better off dead' (FAWC, 2009, p. 14)? I suggest that this is probably a small proportion (although this may represent a large number of suffering animals, given the vast total). This may be true even among intensively reared meat chickens and turkeys, regarded by many people as one of the worst systems for welfare. Webster (1994, p. 156) suggests that: 'Approximately one-quarter of the heavy strains of broiler chicken and turkey are in chronic pain for approximately one-third of their lives.' Yet this means that three-quarters are not in pain (although they may have other welfare problems) and that the rest are free of pain for two-thirds of their lives. This is not to suggest that the pain of broilers and other farm animals is unimportant, far from it, but it does suggest that most farm animals do have a life worth living. If this is so, should we actually try to reduce their number or to increase it? That question may be impossible to answer. There is an extensive discussion in philosophy about whether this sort of approach is valid for human lives – for example, whether we should prefer a higher total sum of welfare in a large population or a higher average in a smaller population (Parfit, 1982) – but there are no clear conclusions for either humans or animals (see Franco *et al.*, Chapter 4, this volume). However, the aspiration that a growing proportion of animals should have a good life (FAWC, 2009) is also relevant here. It might, in theory, be possible to maintain the present numbers of farm animals or even increase them and ensure that many or all have a good life. Given the existing difficulty of avoiding numerous, and often major, welfare problems, though, and also environmental constraints, this possibility is likely to remain theoretical. In practice, achieving a good life for farm animals is more likely if they are kept in smaller numbers and better protected (which would also involve many other changes in the agricultural system, including higher meat prices).

This suggestion, that welfare can best be improved by reducing animal numbers and looking after those animals more carefully, is compatible with increased attention to animals as individuals, stressed as important above. It is also germane to the comparison between farming of relatively small numbers of mammals and large numbers of poultry – particularly chickens and ducks. Of course, there are some systems, in some countries, in which mammals are kept in large numbers, often in barren conditions, such as the beef feedlots of Midwestern USA. But globally, these are dwarfed by broiler farms on which chickens are kept in groups of many thousand, in even more barren conditions. The large group size and intensive approach to poultry farming is associated with the smaller body size of the animals, which both yields less profit per animal and allows more to be kept in the same area. Similarly, among the main farmed mammals, pigs, sheep and goats are kept in greater numbers worldwide than larger-bodied cattle (although cattle are most numerous in many developed countries).

Numbers also have impacts on how animals are perceived. Farmers keeping animals in large groups are less likely to give them individual care and attention (particularly if the economic value of each is small), or perhaps even to think of them as individuals, than if they have more select numbers. It would be interesting to tag some animals in a large group, such as a commercial broiler flock, so that they are individually identifiable, and monitor whether this changes the farmer's attitude to or treatment of both those individuals and others. Numbers probably also affect consumer attitudes. As said above, some people probably empathize less with poultry than with mammals. This may be partly *because* they are in large numbers and difficult to identify with individually.

Welfare is not always worse in large groups than small (see Rushen and de Passillé, Chapter 10, this volume), but large groups (and the barren conditions in which they are usually kept) do often cause problems. And certainly, dealing with large numbers of animals in housing or transport or at a slaughterhouse can lead to neglect, or even cruelty.

So, animal numbers are important. This is not the place to discuss the many factors that affect either farm size or group size within farms, although it should be mentioned that these might be different between farms producing niche products such as organic or high-welfare animal products and those producing for the commodity market. However, one final important factor in consumer choice that contributes to these variables – and one that is considered surprisingly rarely in relation to animal welfare – is the amount of animal products that people eat.

6.8 Amount of Animal Products Eaten

It is well recognized that the average amount of animal protein that people eat varies widely between countries, and in many developed countries is greater than the protein needed nutritionally (Table 6.2). There is also, of course, considerable variation within countries. And this choice both by individual people and cumulatively across society obviously affects the numbers of animals reared and killed. To give one concrete example, Aladfar (2005) estimated that if a person ate one 170 g (6 oz) portion of beef/day, it would take between 1200 and 1333 days to eat all the meat from an average cow. Eating 230 g (8 oz) portions reduces that to 900–1000 days and increases by one-third the number of cows required.

As with the general comments about animal numbers in the previous section, there need be no direct link between meat and milk consumption, animal numbers and animal treatment. However, reducing the numbers of animals through reduced consumption may, in practice, again be associated with them being looked after better, at least when people consume less through choice and not through poverty. This is because a choice to eat less meat or dairy products is likely to be associated with some of the other choices discussed above, considering the effects of diet on health, ecology or animal welfare. There is

Table 6.2. Average consumption of animal protein/person/day in grams. Recommended protein intake is 58 g (FAO, undated).

North America	72
Oceania	68
Northern Europe	62
Eastern Europe	49
South America	42
Eastern Asia	35
Africa	13
South Asia	12

evidence that people who buy organic meat buy less meat (Hoffmann and Spiller, 2010). This does not seem to be just because it costs more, but because they regard it as special, a part of their diet to be savoured rather than a routine commodity. It is true that organic and high-welfare produce does generally cost more than commodity products; this is one reason why at least some farmers can make a living, keeping smaller numbers of animals in better conditions and selling smaller volumes of product. If more people who currently eat more than a minimum amount of animal produce ate less but paid a higher price for it, and considered the provenance of that produce carefully, there could be benefits for themselves, for the animals, for the farmers and for the planet.

6.9 Food with a Name

Much of the discussion comes back to the point raised at the outset, the importance of considering sentient animals as individuals. I said that some of the reasons people choose animal products from certain categories, such as choosing fish or poultry over mammals because they are 'not like us', are irrelevant to this. It may even be contradictory to the welfare of the individuals affected, being associated with large-scale systems in over-simple environments such as those used for most fish and poultry production. Individual care is more likely in large, high-value animals kept in smaller groups. The balance of environmental and welfare arguments suggests that we should keep fewer livestock but treat them better, and as part of that, it may be beneficial to favour larger mammals rather than smaller fish or poultry (despite the fact that production of the former is currently less efficient than the latter in terms of greenhouse gas production per kilogram of meat: see Hötzel, Chapter 5, this volume). Doing so will mean that fewer animals are reared and killed for people to eat, a tendency that will be strengthened if, as one aspect of being conscientious consumers, people reduce their intake of animal produce.

Thus, while it is possible to buy chicken meat from a bird that has had a good life, in practice a chicken reared in a commercial system, even a system designed to promote welfare such as free range, will have lived in a large group in relatively simple conditions. There are grounds to argue that such a life is not as good as that of a cow living in a near-natural social group in a near-natural environment, which the farmer can provide because of the cow's high value. Furthermore, while a single cow will provide around a thousand meals, a single chicken will provide only about six of the same size.

What about veal calves, which are a special category of cattle? Veal has long been associated with poor welfare, because veal calves are often reared in crates so small that they cannot turn around and are fed diets deficient in roughage and iron, to produce tender white meat. Yet veal calves are of high individual financial value, and this has enabled the creation of a market for higher-welfare, pink-coloured veal from animals reared in small groups in much better conditions and fed more natural diets: 'rose veal' (sometimes called 'rosé veal'). Indeed, Rose Veal has become a recognized brand, sometimes labelled by the RSPCA Freedom Food standards (RSPCA, 2011) but sometimes simply connoting a high-value, high-quality, high-welfare category of meat.

So, veal can be better for welfare than chicken.

Here is a proposition: if you care about animal welfare you should choose food with a name.

In other words, an approach that is particularly likely to safeguard welfare is choosing animal products from a system or a farm which keeps animals in small groups and in which the animals are known and cared for individually, to the extent that they are actually – or at least might potentially be – given a name.

Naming is more likely in longer-lived animals than those that are shorter lived (and veal calves, while they have shorter lives than other beef cattle, are still longer lived than broiler chickens; see also Franco *et al.*, Chapter 4, this volume). It is most likely with pigs, sheep, goats and especially cows. It is possible, but unlikely, with chickens. It is probably impossible with fish – not only because they are kept in very large groups of very similar individuals but also because of other issues, such as the sheer difficulty of seeing them underwater.

Some people say that they dislike the idea of 'food with a face'. This may be just another way of saying 'I am a vegetarian', but it may also be an assertion that eating a whole animal, such as a fish complete with its head, is an uncomfortable reminder that this food was once a living being. And it is true that emphasis on the fact that food comes from a particular, individual animal – perhaps even to the extent of knowing its name – may also be uncomfortable, as I said at the outset in relation to the question, 'Whom should we eat?'. But it may nevertheless be better for the animal and, to that extent, reduce another source of discomfort, the worry that the animal providing you with food had a life that was worse than it might have been.

Even if the animal does not have a name, or if you do not want to know it, other names associated with your food are also important, such as the name of the farmer or farm that produced it and the name of the brand under which it is sold. Knowledge of the welfare provenance of your food is important for assurance about how animals have been reared and how their successors will be reared in future, and part of that is knowing who reared them, on which farm and by what marketing criteria: let us say, for example, Harry Rose, Rose Farm or Rose Veal.

6.10 Conclusions

- Some people believe that we should not eat any products from sentient animals, either considering that the most important harm done to animals is killing them or thinking that while it might be justifiable to give animals a good life and a gentle death, it is not possible in practice.

- For those who think that eating food from sentient animals is justifiable, these animals should have the best possible welfare up to and including slaughter.
- Ensuring better than average animal welfare should generally be achieved by selecting food labelled as humane, or with another label or source that provides some evidence of this.
- One factor sometimes used in choice is taxonomic group, such as fish, bird or mammal. This may be relevant to empathy, but animals 'not like us' are not intrinsically likely to have better welfare. Fish and poultry are usually kept in large-scale barren environments that may be unfavourable to welfare.
- Welfare may be best improved by reducing animal numbers and looking after them more carefully, with increased attention to individuals.
- So, some of the most important factors affecting welfare are body size, longevity, group size, the amount of animal products people eat and the price the farmer receives. Farmers keeping large high-value animals in small groups are best able to give them individual care. Choosing products from large animals rather than small also addresses concerns over killing: one cow provides meat for about one thousand meals, a chicken about six.
- It is proposed that to safeguard welfare it is best to choose 'food with a name', from a system or farm in which the animals are treated individually, to the extent that they are – or potentially might be – given a name. Other names associated with food are also important: the name of the farmer or farm that has produced it and the name of the brand under which it is sold.
- Choices maximizing the welfare of farm animals must be balanced against other factors such as health and environmental impacts, but are of particular ethical significance because of the sentience of the animals concerned. Whom should we eat? It is probably easier to ensure that long-lived, high-value cattle (e.g. rose veal calves) have a good life than short-lived, low-value chickens.

Questions for discussion

1. Some people go to a local farm and get to know the pig or the calf from which they will later be supplied meat. Would you be willing to do the same?
2. In some places, meat is sold with a picture of the individual farmer who raised the animal in question. Is this a good idea? Would it also be a good idea to have a picture of the animal?
3. Ignoring environmental arguments, if farm animals could be allowed good lives, should we increase their number (i.e. the number born, living and killed for food)? Is the answer to this different if we argue from our own (human) point of view or from the animals' point of view?

4. What would be an ideal farming system for animal welfare; for example, animals living undisturbed in a near-natural habitat until killed suddenly by shooting? Or animals kept for milk or eggs until dying natural deaths?
5. What are the moral issues involved in eating shellfish, insects and other animals for which consideration of sentience is less important?

Acknowledgements

I am grateful to Lesley Lambert, Claire Bass and all the other authors of this book, especially Ngaio Beausoleil, for helpful comments on earlier drafts of this chapter.

References

Aladfar (2005) Contributor to: How much meat is there on a cow? (ask.metafilter.com/27259/How-much-meat-is-there-on-a-cow, accessed 18 February 2013).

Anon. (2011) Chart: How much meat is on a blue whale? (www.good.is/post/chart-how-big-are-blue-whales-1/, accessed 18 February 2013).

Appleby, M.C. (1999) *What Should We Do About Animal Welfare?* Blackwells, Oxford.

Appleby, M.C. (2003) The European Union ban on conventional cages for laying hens: history and prospects. *Journal of Applied Animal Welfare Science* 6, 103–121.

Braithwaite, V. (2010) *Do Fish Feel Pain?* Oxford University Press, Oxford.

Branson, E.J. (ed.) (2008) *Fish Welfare*. Blackwells, Oxford.

Davis, S.L. (2003) The Least Harm Principle may require that humans consume a diet containing large herbivores, not a vegan diet. *Journal of Agricultural and Environmental Ethics* 16, 387–394.

FAO (Food and Agriculture Organization of the United Nations) (undated) FAOSTAT (www.fao.stat.org, accessed 18 February 2013).

FAWC (Farm Animal Welfare Council) (2009) *Farm Animal Welfare in Great Britain: Past, Present and Future*. Defra, London.

Foer, J.S. (2009) *Eating Animals*. Little, Brown and Co, London.

Francione, G.L. and Garner, R. (2010) *The Animal Rights Debate: Abolition or Regulation*. Columbia University Press, New York.

Harrison, R. (1964) *Animal Machines: The New Factory Farming Industry*. Vincent Stuart, London.

Haynes, R.P. (2011) Competing conceptions of animal welfare and their ethical implications for the treatment of non-human animals. *Acta Biotheoretica* 59, 105–120.

Hearne, V. (1982) *Adam's Task: Calling Animals by Name*. Vintage, New York.

Hill, J.L. (1996) *The Case for Vegetarianism: Philosophy for a Small Planet*. Rowman and Littlefield, Maryland.

Hoffmann, I. and Spiller, A. (2010) Auswertung der Daten der Nationalen Verzehrsstudie II (NVS II): eine integrierte verhaltens- und lebensstilbasierte Analyse des Bio-Konsums [Data Interpretation Based on the German National Nutrition Survey II (NVS II): An Integrative Analysis of Behavioural and Lifestyle-Related Factors for Organic Food Consumption] (http://orgprints.org/18055/1/18055-08OE056_08OE069-MRI_uni-goettingen-hoffmann_spiller-2010-verzehrsstudie.pdf, accessed 18 February 2013).

Matheny, J.G. (2003) Least harm: a defense of vegetarianism from Steven Davis's omnivorous proposal. *Journal of Agricultural and Environmental Ethics* 16, 505–511.

Mood, A. (2010) Worse things happen at sea: the welfare of wild-caught fish (www.fishcount.org.uk/published/standard/fishcountfullrptSR.pdf, accessed 18 February 2013).

Mood, A. and Brooke, P. (2012) *Estimating the Number of Farmed Fish Killed in Global Aquaculture Each Year*. Compassion in World Farming, Godalming, UK.

Newkirk, I. (undated) (www.discoverthenetworks.org/individualProfile.asp?indid=2072, accessed 7 June 2013).

Norwood, F.B. and Lusk, J.L. (2011) *Compassion by the Pound: The Economics of Farm Animal Welfare*. Oxford University Press, Oxford, UK.

Olsson, A. (2008) Grow meat, but off the bone. *New Scientist* 5 (July), 18.

Parfit, D. (1982) Future generations: further problems. *Philosophy and Public Affairs* 11, 113–172.

Pollan, M. (2006) *The Omnivore's Dilemma: A Natural History of Four Meals*. Penguin, New York.

Roth, B. and Øines, S. (2010) Stunning and killing of edible crabs (*Cancer pagurus*). *Animal Welfare* 19, 287–294.

RSPCA (2011) Freedom food (www.rspca.org.uk/freedomfood, accessed 18 February 2013).

Singer, P. and Mason, J. (2006) *The Ethics of What We Eat: Why Our Food Choices Matter*. Text Publishing Company, Melbourne.

Webster, A.J.F. (1994) *Animal Welfare: A Cool Eye Towards Eden*. Blackwell, Oxford.

7

Public Health and Animal Welfare

Carla Forte Maiolino Molento*
Animal Welfare Laboratory, Federal University of Paraná, Brazil

7.1 Abstract

What are the links among the concepts of global public health, human health and animal welfare? The objective of this chapter is to discuss the relationship between public health and animal welfare issues, from the historical perspective of rabies control to the more recent challenges brought by the populations of different vertebrate species in urban environments. From what has been learned so far, considering both the efficiency and ethics of animal population control strategies, a change in paradigm is needed. The killing of animals for the sake of public health appears to be ineffective in the case of dogs and other animals studied; at this stage, proponents of this strategy should carry the burden of proof for its effectiveness based on knowledge of the population dynamics of the species involved. In some cases, urban animals can play a positive role in the control of zoonotic disease problems, especially if they receive health and welfare attention. Overall, zoonotic diseases, as well as health and welfare maintenance, occur within dynamic processes which cannot be dealt with by employing only short-term approaches.

7.2 Introduction

Global public health is 'the collective action we take worldwide for improving health and health equity, aiming to bring the best available cost-effective and

*E-mail: carlamolento@ufpr.br

feasible interventions to all populations and selected high-risk groups' (Beaglehole and Bonita, 2008, p. 1988). Health itself is defined by the World Health Organization as 'a state of complete physical, mental, and social well-being and not merely the absence of disease or infirmity' (WHO, 2012a). This definition of human health is grounded on the idea of well-being, a synonym for welfare, as usually used in the case of animals. One concept of animal welfare is the state of an individual as regards its attempts to cope with its environment (Broom and Fraser, 2007); others have defined animal welfare as a concept related basically to animal feelings (Duncan and Petherick, 1991). An important idea behind the animal welfare concept is to look for an understanding of the animal's life quality from the perspective of the animal and take appropriate action according to such understanding. What are the links among these concepts?

The objective of this chapter is to discuss the relationship between public health and animal welfare, from the historical perspective of rabies control to the more recent challenges brought by the populations of different vertebrate species in urban environments. It is important to note that the approach used here relies on the understanding of vertebrates as sentient beings, who strive for both quality and quantity of life, at least in the sense that they all strive to stay well and alive. The literature review presented here was motivated by the need for a better world for both people and animals.

7.3 Public Health and Animal Welfare Issues

Humans and other animals share the same world, similar basic needs and, if we consider vertebrates, similar basic feelings, as concluded recently by a prominent international group of scientists (The Cambridge Declaration on Consciousness, 2012). We also share many of the same basic problems, such as hunger, cold and disease.

Of the 1425 known human pathogens, 61% are zoonotic, and within the last generation, around 40 new human diseases have been discovered, 75% of them zoonotic in character (Davis, 2011). A quick glance at these numbers and it becomes evident that we cannot move forward in finding solutions for the improvement of human health if we look only at humans. This relationship between humans, animals and diseases has been known for centuries and is the reason behind many human interventions in animal life.

Most of the literature tends to employ terminology that implies that animals are the source of disease and humans the victims. Actually, the number one carrier of human infectious diseases is probably humans themselves, as are dogs for dogs, chickens for chickens, and so on. This is illustrated throughout human history, as shown by an example taken from McNeill (1998). During the Spanish conquest of Mexico, why did relatively few Spaniards succeed in impressing their culture on an enormously larger number of Amerindians?

For 4 months after the Aztecs had driven Cortez and his men from their city, an epidemic of smallpox raged among them. Such an epidemic, striking an entirely inexperienced population, killed something like a quarter to a third of the Aztec people. The arrival of a new pathogen from Europe was devastating. We can probably derive a general rule, even more relevant for modern situations, according to which the more individuals move around, the easier it is for infectious diseases to spread. This rule probably remains true when we interfere with animal populations, increasing their turnover. Fortunately, we have learned a great deal about how to reduce disease transmission among humans, including the use of vaccines and the development and application of sanitary measures.

Whenever there is disease, all affected are victims, and the suggestion that specific diseases have been brought about by animals is sometimes unproven. As an example, some emerging zoonotic diseases are even named after animals, such as mad cow disease and bird and swine flu. However, according to Graham *et al.* (2008), these zoonotic diseases may result from relatively recent human practices of intensive confinement or feeding animals food that the animals themselves would never choose to eat. Superficially, animals may be considered the origin of these diseases. However, a deeper analysis often shows that human practices share some responsibility for these outbreaks.

As discussed by Davis (2011), our image of public health today should be radically different from what it was at the turn of the 20th century. In 1900, infectious diseases topped the list of causes of death (Table 7.1). According to the World Health Organization (WHO, 2008), by 2030 there will be only two infectious causes among a list of chronic, mostly preventable diseases.

Table 7.1. Top ten causes of death in humans at the turn of the 20th century and in 2030, as estimated by the WHO (2008), with infectious diseases in italics. Modified after Davis (2011).

Ranking position	Year	
	1900	2030
1	*Pneumonia*	Ischaemic heart disease
2	*Tuberculosis*	Cerebrovascular disease
3	*Diarrhoea/enteritis*	Chronic obstructive pulmonary disease
4	Heart disease	*Lower respiratory infections*
5	Stroke	Road traffic accidents
6	Liver disease	Trachea, bronchus, lung cancers
7	Injuries	Diabetes mellitus
8	Cancer	Hypertensive heart disease
9	Senility	Stomach cancer
10	*Diphtheria*	*HIV/AIDS*

Sadly, there are issues of public health inequity throughout the world, and the top ten list varies not only across centuries but also across countries. We may also consider the importance of infectious diseases when understood within the main health problems that humanity faces. What are the most difficult obstacles to human health and human life worldwide? How meaningful is the decrease in infectious diseases within the top ten causes of death in Table 7.1? There are difficulties in obtaining reliable data from the whole world but, to the best of our knowledge, infectious diseases remain important (Table 7.2) and efforts to prevent them represent a good use of our resources. In the countries where the percentages of all death causes due to the categories presented in Table 7.2 are relatively low, namely the USA and Germany, the shift made towards more chronic diseases indicated in Table 7.1 has already taken place. In South Africa, it seems that infectious and parasitic diseases still top the list of human death causes. The Brazilian percentage of deaths due to interpersonal violence is high, and this may be common to other countries as well. Thus, ideal public health actions and programmes should promote respect for the needs of others.

As a concrete example of the relationship between public health and animal welfare issues, consider humans and dogs. Dogs outside of responsible guardianship bring many public health risks. Zoonotic diseases are a primary concern. Dogs live close to humans and share a biology that is similar to ours in many ways. Diseases transmissible between dogs and humans are numerous and may be viral (such as rabies), bacterial (leptospirosis), parasitic (roundworms, fleas), fungal (ringworm) or protozoan (leishmaniosis). There are impacts of stray dogs on human health which are additional to the shared disease problems. Dogs may be victims of road accidents, which cause human suffering as well. Road accidents are the major contributor to the category of accidents and adverse effects seen in Table 7.2. Stray dogs produce faeces and urine in large quantities, due to their large population size, creating a source of environmental pollution. Stray dogs also may be a source of noise through excessive barking, and may bite people or other animals.

A welfare problem in relation to dog genetics that overlaps with public health concerns is the selection for aggression. In general, dogs have a higher threshold for aggression than wolves (Stafford, 2006), but some aggressiveness remains. Aggression tops the list of behavioural problems, both as reported by owners and as diagnosed in behavioural clinics (Stafford, 2006). Some breeds of dog have also been selected for aggressiveness. Aggressive dogs can end up on the street. For example, in Curitiba, Brazil, after the approval of a law making guard dog rental a crime (Brasil, 2008), many Rottweilers were abandoned on the streets. Questions related to breeding for aggressiveness should be included as public health concerns for reasons of human and dog welfare. There are many dog-to-dog and dog-to-other-animal attacks: aggression is a frequent and important problem between families and their pets, which may lead to abandonment, and aggressive dogs may end up killed rather

Table 7.2. Number of citizens within each category of cause of death, and human population size, according to the World Health Organization (WHO, 2012b) in 2004, and Human Development Index and life expectancy at birth, according to the United Nations Development Programme (UNDP, 2012) in selected countries.

Selected category of cause of death	USA		Germany		Brazil		South Africa	
	Human deaths	Per cent of all causes of death (‰ total population)	Human deaths	Per cent of all causes of death (‰ total population)	Human deaths	Per cent of all causes of death (‰ total population)	Human deaths	Per cent of all causes of death (‰ total population)
Homicide and injury purposely inflicted by other people	17,165	0.72 (0.058)	526	0.06 (0.006)	48,349	4.72 (0.262)	4,216	0.74 (0.089)
Suicide and self-inflicted injury	32,363	1.35 (0.110)	10,733	1.31 (0.130)	8,015	0.78 (0.043)	387	0.07 (0.008)
Accidents and adverse effects	114,895	4.79 (0.391)	19,458	2.38 (0.236)	58,791	5.74 (0.319)	11,513	2.03 (0.242)
Infectious and parasitic diseases	64,618	2.70 (0.220)	11,062	1.35 (0.134)	46,049	4.50 (0.250)	132,235	23.37 (2.781)
All causes listed on the WHO (2012b) website	2,397,615	100.00 (8.165)	818,271	100.00 (9.918)	1,023,372	100.00 (5.552)	565,953	100.00 (11.905)
Total population	293,655,404		82,501,274		184,317,693		47,540,925	
Human Development Index (life expectancy at birth, years)	4 (78.5)		9 (80.4)		84 (73.5)		123 (68.8)	

than treated, due to the risks of rehoming. Aggression tops the list of behavioural problems, both as reported by owners and as diagnosed in behavioural clinics (Stafford, 2006). It is relevant from the public health and the animal welfare points of view that dog aggressiveness be prevented both by genetics and environmental strategies; in other words, both in living dogs and in those yet to come.

For the reasons given above, many interventions have the goal of controlling dog populations. These interventions typically used to be planned with an anthropocentric view: a single focus on humans was a general characteristic, in tune with the predominant paradigm across most centuries of occidental culture. The practice did bring improvements in human life, and probably represented the best that could be done with the knowledge and the understanding then available. Perhaps the time has come to change this paradigm if we are to do the best for people and animals with the knowledge and the understanding available today.

7.4 History of Dog Population Control in the Context of Human Disease Control

The main motivation to control dog populations was human health, as has been evidenced by the leadership of the World Health Organization on this issue since the 1960s (WHO, 2005). The history of urban dog control is intertwined with the history of rabies control. Dogs have been a major factor for human rabies, both because they have long been in close association with humans and because they are able to bite. Until rabies vaccines became available for large-scale use in animals, the authorities charged with human disease control then employed the rationale that controlling the dog population would help decrease the risk of disease. So, emphasis was put on rabies control through dog population control. Many countries launched programmes for dog elimination. These included dog clubbing, drowning, shooting, elimination using gas and vacuum chambers, poisoning and electrical killing. Gas chambers were developed to provide a more humane method for dog killing as compared to clubbing or drowning, and the use of decompression chambers was introduced in the 1970s (Zawistowski and Morris, 2004). The chambers also provided a less distressing method for the people. Zawistowski and Morris (2004) report that carbon monoxide is still used to kill dogs in a number of animal shelters in the USA.

Simultaneous to the efforts on dog population control, rabies vaccination became increasingly available. During the period 1993–2002, the countries of the Americas reported a decrease of 82% in the number of human rabies cases (Belotto *et al.*, 2005). This sharp reduction was attributable mainly to the mass vaccination of dogs and prophylactic treatment for people who had been exposed (Belotto *et al.*, 2005). These data suggest that the control of rabies is related more specifically to the control of dog immunity than to the control of dog population size. Although both types of control may have potential positive impacts on dog welfare, the strategies historically employed to control dog numbers were often related to severe direct and indirect negative impacts on dog welfare, ranging from aversive methods of capture and handling, very poor welfare during maintenance in overcrowded and low-resource environments and the inhumane methods for dog killing described above, and all portray a negative interaction between humans and stray dogs.

In addition to the negative effects of capturing and killing in terms of animal welfare, research has begun to show that dog capture actually increases the population turnover rate, rather than decreasing the population size, in geographic areas where birth rates are high. Higher population turnover and higher population flow through movement are coupled to increases in contagious diseases. This issue was recognized by the World Health Organization expert consultation on rabies (WHO, 2005). The report emphasized the inefficacy of dog capture and killing within the efforts to control rabies.

One difficulty in monitoring the efficacy of control programmes is the lack of knowledge of population size. Recently, efforts have been put into the estimation of dog population size in different regions of the world, such as the USA (Beck, 2002), India (Hiby *et al.*, 2011) and South America (Fig. 7.1). In Latin America, estimates in 2001 put the dog population size at close to 65 million, half of it in areas where rabies was present (Belotto *et al.*, 2005). Data produced in South America (Fig. 7.1) suggest that the dog population size might be significantly larger. Taking locally produced information of an average dog to human index of 1:4, the estimated dog population size for Brazil alone is 47,700,000, as indicated by its 190,700,000 human inhabitants (IBGE, 2010).

Consider the city of Curitiba (State of Paraná, South Brazil) as an example. The city was the first in Brazil to be declared free of canine rabies. This success was due to the efforts on dog rabies vaccination (Schneider *et al.*, 1996), as control of the size of the dog population had not been effective. The difficulty in controlling population size becomes obvious when considering the population dynamics. Taking again a dog to human index of 1:4, Curitiba, with its 1,750,000 human inhabitants (IBGE, 2010), has an estimated population of around 440,000 dogs. Approximately half of this population is female. Assuming, for example, 25% responsible pet guardianship and two-thirds of the free-roaming bitches being reproductively active, there are approximately 110,000 reproductive bitches with free access to the streets. If each of them produces an average of four puppies per litter and they average two litters per year, the total number of puppies produced per year is around 880,000. Thus, any serious effort to control population size must involve killing or sterilizing hundreds of thousands of dogs.

Localization on the map	Reference	Dog to human ratio
1 Brazil, State of Paraná, Curitiba	Damasco *et al.*, 2005	1:3.3
2 Brazil, State of Paraná, ten rural villages	Molento *et al.*, 2007	1:2.7
3 Brazil, State of Paraná, Araucária	F.M. Wolff, Araucária, 2005, personal communication	1:4.0
4 Brazil, State of São Paulo, Taboão da Serra	Dias *et al.*, 2004	1:5.1
5 Brazil, State of São Paulo, Araçatuba	Nunes *et al.*, 1997	1:3.6
6 Brazil, State of São Paulo, Serra Azul	Matos *et al.*, 2002	1:5.0
7 Brazil, State of São Paulo, Ibiúna	Soto, 2003	1:3.8
8 Brazil, State of Minas Gerais, Ouro Preto	Naveda *et al.*, 2002	1:2.6
9 Argentina, Almirante Brown	M. Antoniazzi, Belo Horizonte, 2005, personal communication	1:4.0
10 Venezuela, Mucuhíes	Coppinger and Coppinger, 1998	1:2.5
11 Chile, Viña del Mar	Morales *et al.*, 2009	1:4.1

Fig. 7.1. Dog to human ratios in 11 South American locations.

The government in Curitiba employed a dog capture-and-kill strategy for three decades, which ended in November 2005. During the last decade of the programme, the numbers varied from 10,000 to 15,000 dogs killed per year (Biondo *et al.*, 2007). Despite its costs in terms of money and human and animal suffering, it served no purpose for dog population control, as attested by the simple calculation above and by a walk around the city (personal observation). Two conclusions may be drawn: (i) the best course of action to control the dog population in Curitiba effectively is through education towards responsible pet guardianship, which would involve more than a million citizens in taking care of hundreds of thousands of dogs; and (ii) the capture-and-kill strategy is a useless effort and a misuse of public money. Capturing stray dogs and taking them to shelters is likely to be no better. As Zawistowski and Morris (2004, p. 3) commented: 'Animal shelters have long occupied a spot on the periphery of a community's geography and awareness. They have been the place where what we don't want to think about happens to pets we no longer care about.'

More detailed modelling of dog population dynamics has been developed, with inputs derived from Brazilian data (Amaku *et al.*, 2009). The estimated outcomes point to the same conclusions described above. Dog population size is resistant to human efforts of control. Even if a large proportion of the population were killed or sterilized each year, a decrease in total population size would take years to achieve (Fig. 7.2). Thus, in regions that face similar constraints, there are no feasible short-term means to reduce stray dog populations. The important steps towards responsible pet guardianship must be taken, coupled to adjusted expectations for results in a medium- or long-term timeframe. Meanwhile, there is a need to accept the presence of stray dogs and act to decrease the likelihood of the public health problems they may present, as well as the animal welfare problems they may face.

Some countries have made dog killing a more permanent option for their citizens, such as some areas in the USA. In some locations, such as Baltimore (Beck, 2002), citizens may call the animal shelter to remove pets they no longer want. Other countries, such as Brazil, have employed dog elimination strategies for decades and never achieved dog population control. The different results obtained are probably due to cultural, social, economic and even geographical context. Environmental conditions, such as a relatively warm climate, the presence of food in garbage dumps and offered by people, and the migration from neighbouring dog populations may be favourable to the maintenance of stray dog populations. Another important factor is the degree of socio-economic development: people lacking basic life conditions may not be able to provide responsible pet guardianship, even if they are willing to do so. Different levels of acceptance to dog killing also exist. All these factors contribute to the need for local approaches to the issue of dog population control. As concluded by Palmer for cats (Chapter 9, this volume), the best solutions will be those designed specifically for each context.

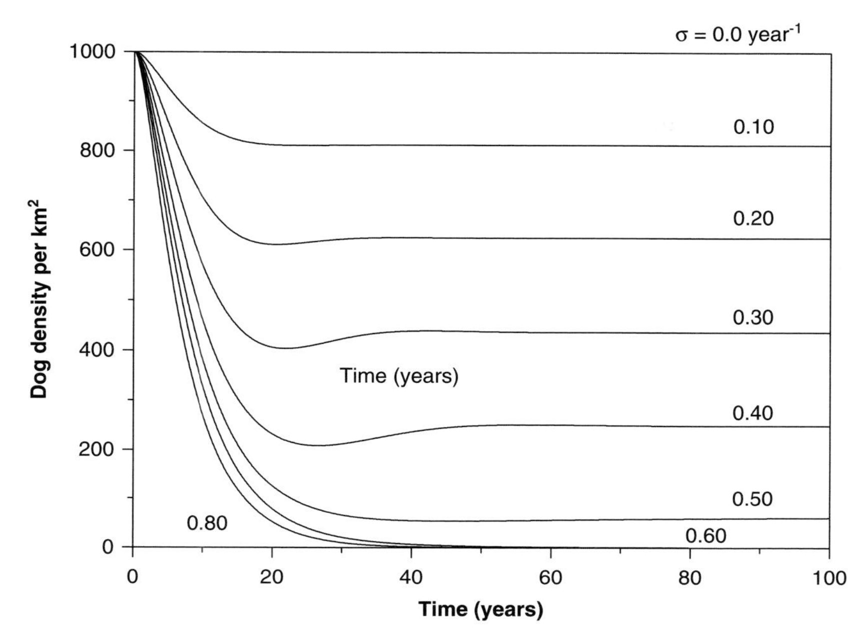

Fig. 7.2. Hypothetical dog population and the potential effects of different sterilization rates, assuming no migration to the area ($\sigma = 0.0$); simulation shows that even for rates as high as 80% (lowest slope), it would take around 5 years for a noticeable change in population size, say a reduction of 50%. Adapted from Amaku *et al.*, 2009.

7.5 New Strategies for Dog Population Control

Ideally, there would be no stray dogs, for both human and dog welfare reasons; in this sense, the dilemma between public health and animal welfare is false. However, as may be deduced from the earlier section, a strategy to achieve an immediate absence of stray dogs is lacking in many regions of the world. Is it possible to decrease the likelihood of the public health problems when stray dogs are around? Yes it is, as demonstrated by the history of dog rabies control. The more care given to stray dogs, the less likely they are to become sick. If a proportion of this population were to be sterilized, it would produce benefits from both a direct perspective, as they would not produce the puppies that would maintain the stray population, and an indirect one, helping education towards a respectful relationship that, in turn, might contribute to the establishment of responsible pet guardianship. The intervention proposed does not involve the removal of the animal from the neighbourhood, but is framed as capture–spay/neuter–release or capture–vasectomize–release (Molento, 2004).

The longer these sterilized, healthy stray dogs live, the longer the protection against dog diseases in the area, as their presence decreases the chances of foreign dogs moving in. Foreign dogs may bring pathogens and are likely to be reproductively active. Thus, stray dogs that are under a certain level of control may be understood as a sanitary and a reproductive barrier for the communities where they live. In other words, it seems that keeping healthy and sterilized stray dogs and fostering an increase in their life expectancy represent the best that can be done for the sake of both public health and animal welfare, until responsible pet guardianship is a reality. The evidence for these claims specifically for dogs is yet to be produced, but data from other urban species such as pigeons (see below) and from wild animals suggest that this rationale of filling the environmental carrying capacity might be applicable for dogs, except perhaps in scenarios where no surplus young animals or adults are available.

The stray animals that become part of a public health and animal welfare programme have been called community dogs in Paraná (Brazil). They are identified, vaccinated, dewormed, surgically sterilized and then returned to their community. A local dweller accepts the responsibility as the dog's caregiver. His or her responsibilities are to feed the animal (which they normally do already) and to call the disease control centre if anything happens to the animal. The disease control centre maintains periodic contact with all caregivers. This work has brought an unexpected level of contact between animal control agents and the community. Beck (2002) reports observing people harassing the dogcatchers by cursing them and by chasing dogs away. This report is consistent with the situation in many Brazilian cities, where people may additionally hide stray animals from the service, as reported by some health officials in the State of Paraná. A significant shift occurred in the relationship between disease control centres and the communities involved when the community dog pilot project started. The willingness of people to collaborate became higher and the relationship turned more positive (Fig. 7.3).

Community dogs already exist in many societies. A study in Baltimore, USA, in the early 1970s reported that these dogs were often not truly owned by one person or one family but were sometimes fed and protected by individuals in an area (Beck, 2002). The proposal of educating society not to feed hungry dogs is not acceptable from an animal welfare point of view. We should not down-regulate compassion; it is something to be fostered, not restricted. Additionally, feeding stray dogs may be of benefit to public health. Free-ranging dogs that were fed by residents showed a smaller home range than free-ranging dogs not fed by residents (Beck, 2002); this might present advantages from the perspective of contagious disease prevention. Besides, dogs that are better nourished are at lower risk of disease, presenting higher immunological defences. In southern Brazil, it is common to see food and shelter offered to street dogs, often by families that already have their own dogs. Beck (2002) commented on a free-ranging dog called 'Shaggy' that had a good body condition score and was considered in good health by a veterinarian, except for

Fig. 7.3. (a) Aposentado, one of the dogs in the pilot project on community dogs; (b) a maintainer helping animal control officers to take a community dog for surgical sterilization and other health procedures; and (c) the interaction between Feio, another dog in the pilot project, and the local people. Town of Araucária, State of Paraná, South of Brazil. 2009. Photos by Luciana Vargas, reproduced with permission.

(Continued)

Fig. 7.3. (*Continued*)

worms and ear mites, thus presenting an example of a stray animal with relatively good welfare. As a last comment on the community dog programme, it is essential that shelter be provided intentionally. Planned sheltering is more permanent and adequate for the dog, and also more adequate for the neighbourhood, in the sense that disturbances may be avoided. For instance, shelter may be moved away from main roads, thus decreasing risks of a car accident.

Strategies for dog population control must address the rapid turnover of dog populations, one of the major obstacles to the prevention of zoonotic diseases (WHO, 2005). One concrete example of turnover rates comes from the work in the north-west region of the State of Paraná: 3 years after an initial intervention for sterilization and education, only 21% of the dogs were still there (Molento *et al.*, 2007).The rapid population turnover is related to high mortality and low mean age. A younger population is more susceptible to diseases such as rabies and distemper (Beck, 2002). By decreasing birth and death rates and decreasing migration, it should be possible to increase the average population age.

When the population is known, it becomes possible to establish concrete goals and to monitor the effects of any given intervention. It is now possible to say that dog distribution across cities is not even, both in terms of total numbers and within the different subpopulations of free-ranging and of responsibly cared for dogs. Free-ranging dogs were more common in the southern half of the city of Baltimore, which had smaller dwellings compared to the north (Beck, 2002). In Curitiba, in neighbourhoods with higher economic incomes, more than 90% of the dogs were kept under responsible guardianship, while in peripheral neighbourhoods, with much lower incomes, virtually all dogs were free ranging (Biondo *et al.*, 2007). The general pattern seems to be that lower income areas have more stray dogs and wealthier areas have fewer. It is possible to hypothesize reasons and correlations for the higher number of strays in poorer areas (Beck, 2002; Biondo *et al.*, 2007). An important message to convey here is that dog and human welfare are intrinsically connected. This idea is related to the 'One Health – One Welfare' paradigm discussed below.

Maybe this community dog strategy redefines the role stray dogs play in relation to public health. Coppinger and Coppinger (1998) proposed that stray animals were commensal with humans. There may not be any net benefit for people to having stray dogs foraging in the human environment, characterizing the one-sided advantages typical of commensalism. However, in regions struggling to achieve responsible pet guardianship, when some level of care is given to the stray animals, they may become a sanitary and reproductive barrier which favours public health.

Some further steps may also be promising. The development of comprehensive infection control for pets, pet policies and surveillance plans and family care during the selection of pets are some examples (Serpell, 1995). The control of dog commerce is also promising. Ideally, dog reproduction should exist always within a planned context. However, there were approximately 0.55 million unplanned dog litters in the USA in 1996 (Scarlett, 2004), suggesting that neutering both sheltered and privately owned animals before puberty should be a national goal. A norm on the need for sterilization before puppies are sold or donated would be a major step, probably relevant to all countries striving to achieve dog population control (but see Sandøe *et al.*, Chapter 3, this volume, on the obesity and health risks associated with the sterilization of dogs).

The One Health Initiative (2012) is a worldwide strategy for expanding interdisciplinary collaborations and communications in all aspects of health care for humans, animals and the environment. One Health is dedicated to improving the lives of all species and, as such, resonates with many of the themes discussed above. During the last century, human health was largely considered the domain of only human health professionals, animal health a matter for veterinarians and environmental health a matter for environmentalists, with little communication between the fields. Problems were largely viewed from a single perspective only and managed accordingly. This is not, however, a realistic representation of our world, where systems affect one another (Lamielle, 2010). This new approach is welcome, as interactions will receive more attention. The current concept of human health does include human welfare, but this is not always the case for animal health. Thus, the author agrees with Mills's (2012) suggestion that the one health concept now be extended beyond physical health and embraces the concept of one welfare (Fig. 7.4).

As argued above, killing dogs is not always the best approach for reducing health risks to the human population. Perhaps, as proposed by Appleby and Huertas (2011), good management sometimes involves a decision not to take the obvious course of killing animals. Examples in other vertebrate species will now be examined to see what can be learned from them.

7.6 Population Control of Other Animal Species

From pets to pests – the addition of a single letter turns the relationship upside down. As described by Miklósi (2007), Australian dingoes used to be pets,

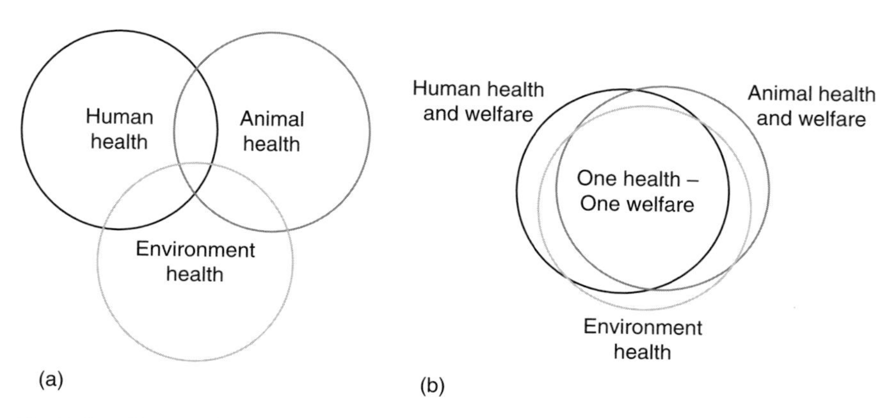

Fig. 7.4. The overlapping areas across human health, animal health and environment health. (a) Traditional perspective on public health; (b) one health – one welfare approach. Modified from Lamielle (2010) to incorporate welfare in the case of sentient beings.

among other relationship formats, to the Aboriginal people before European colonizers brought the domestic dog to the island. The native people then became more prone to keep domestic dogs. In time, dingoes were put on the list of pests to be eradicated. But from an animal welfare perspective, the human perspective on the relationship does not matter. The needs of a rabbit remain the same, regardless of being considered by humans a pet, a laboratory animal, a meat animal or a pest (Broom and Fraser, 2007).

Many vertebrate species live in or near our urban environments and are pertinent to humans for several reasons. For example, Beck (2002) states that they provide insight into the effects of urbanization on humans. Once their ecology is understood, urban dogs may serve as indicators of stress, pollution and environmental deterioration, and as models for behavioural adaptations to urban life. They may also serve as epidemiological indicators, acting as sentinels for human health risks. This concept has advanced within the domain of comparative medicine to the point that, today, this type of field investigation is thought to present some advantages in relation to laboratory research, and to work in both ways: humans may also serve as sentinels to animal health risks (Rabinowitz *et al.*, 2010). This type of strategy may act as a virtuous cycle, with increasing improvements in human and animal health that may include factors of life quality beyond absence of disease. As stated by Beck (2002, p. x): 'Ownerless dogs, living out their lives as wild canids sharing the urban ecosystem with us, are worthy of further study, for they illuminate for us yet another aspect of the animal world.'

In 2008, I was invited to participate in a seminar hosted by the Health Secretariat for the State of Paraná to discuss whether killing pigeons should be exempt from the Animal Cruelty Act (Federal Law of Environmental Crimes), due to public health reasons. The city of Londrina (Brazil) was facing a growing population of pigeons and a recent human death, apparently from *Cryptococcus neoformans*, which may have been transmitted from pigeon excrement. Further analysis showed that the infection was *Cryptococcus gatii*, which is transmitted mostly by inhalation of airborne plant material, but local newspapers were quick to print the headline 'Disease caused by pigeons kills a man'. The alarm eventually abated but the problem remained: can killing the birds help reduce populations and the risk to humans?

A research group in Europe has provided useful information on pigeon population control (Fig. 7.5). The Landelijke Werkgroep Duivenoverlast, or European Working Group for Effective Pigeon Control, stated that:

> The Landelijke Werkgroep Duivenoverlast is keen to stress that the capture and killing of pigeons does not lead to a structural alleviation of damage and nuisance caused by pigeons. Capturing and killing pigeons just creates space that will quickly be taken up by new animals. Because the remaining animals are driven to reproduce more rapidly after a capture 'raid', the population may even grow to a higher level than before.
>
> (Landelijke Werkgroep Duivenoverlast, 2006)

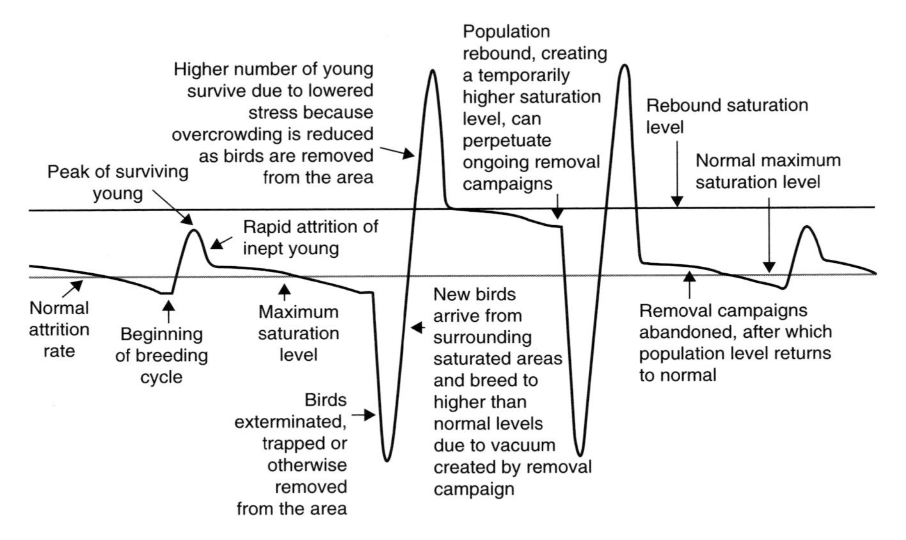

Fig. 7.5. The population cycle of urban pigeons. (From Landelijke Werkgroep Duivenoverlast, 2006.)

Thus, this case resembles the situation with dogs: killing is inefficient at best and more likely counterproductive for population control.

What, then, caused the increase in the pigeon population in Londrina? The State of Paraná is known for grain production. According to Bucher and Ranvaud (2006), an increase in grain availability and waste caused by poor harvesting practices was probably the root of the pigeon population increases. At this point, a simple thought experiment might be of value. We know that population control will not be achieved through the killing of pigeons. However, let us assume that pigeon population control could be achieved this way. What would we have then? The leftover grain would no longer be eaten by pigeons, so other species would probably fill in the empty space left by the decrease in the pigeon population. We may then have an explosion of rodent populations, which would not be a great benefit in terms of public health. Another course of action would be to increase the efficiency of the grain harvest process. This might increase food availability for humans, naturally reduce the pigeon population and not involve using human energy, resources and thinking geared towards killing animals.

The case with pigeons seems to reinforce the need for a change in paradigm within public health actions related to zoonotic disease control. This change has started, for instance by the One Health Initiative, but more efforts are required, as evidenced by the maintenance of dog killing strategies in many countries in the world, or of badger killing, pigeon killing, and so on.

Another example is given by a report on *Nature News* (Gilbert, 2011): Andrew Cunningham, a wildlife epidemiologist at the Institute of Zoology in

London, and his colleagues fear that the next big epidemic could come from henipaviruses, which can cause fatal disease in humans. There is no vaccination to protect against henipavirus. 'We are concerned the solution will be to just kill the bats to control the virus', says Cunningham in the online *Nature News* issue. 'We need to find another way that protects bats and people at the same time.' Bringing the discussion on bats and public health to the present, we may consider the fact that some bat species are today regarded as the major risk for dog, cat and human rabies in Brazil. The development of a strategy of vaccination might be an ideal approach. We were able to eliminate dog rabies using vaccine. We were able to eradicate smallpox from the world using vaccine.

Today we know, as emphasized by Davis (2011), that for understanding of the full scope and meaning of health, for ideal prevention and utilization of health resources and for efficient education and provision of the best treatment procedures there is good reason to expand our focus beyond humans to animals and our environment. This inclusion of animals is not restricted to domestic, native or any other human-centred animal category.

7.7 Challenges Ahead

There are challenges specific to dog population control. As Beck (2002) explains, the sterilizing programme may be misinterpreted as a licence to let pets run free. Community dog programmes also run the risk of being used by people to get veterinary attention for their own animals, simply by letting them go free on the streets. This is a major challenge and a solution will depend on advances in programme implementation and subsequent developments. Another difficulty is that results are not immediate, so there is a challenge in keeping up efforts while little is seen in population level effects. This may be especially problematic for government-run programmes, where the benefits will not necessarily become apparent during the term that politicians are in office. Creating and communicating realistic expectations for programmes will be helpful.

Any strategy adopted to control diseases that are shared between humans and animals will present scientific and ethical challenges. It is to be hoped that we will become more versed in ethical thinking. This, together with the increasing amount of scientific information available, should be instrumental in helping us make a better world with less disease and suffering for all, humans and other animals.

7.8 Conclusions

- The killing of animals for the sake of public health appears to be ineffective in the case of dogs and other animals studied; at this stage, proponents of this strategy should carry the burden of proof for its effectiveness based on knowledge of the population dynamics of the species involved.

- In some cases, urban animals can play a positive role in the control of zoonotic disease problems, especially if they receive health and welfare attention.
- Zoonotic diseases, as well as health and welfare maintenance, occur within dynamic processes which cannot be dealt with by employing narrow, short-term approaches.

Questions for discussion

1. A stray dog or a stray cat regularly visits your garden that also serves as a playground for your children or grandchildren. What would you try to do about this?
2. What would you recommend as the best methods of population control for dogs? What about for cats (see also Chapter 9, this volume)? Might the methods differ for the two species?
3. A local farmer considers starlings to be a pest on his farm and suggests shooting birds. What questions would you ask to determine whether this approach would be effective as a long-term method to control the number of birds on this farm and whether it would be ethically appropriate?
4. This chapter discusses how good animal health can contribute to human health and vice versa. How might concern for environmental health contribute to the animal and human health examples discussed above?

Acknowledgements

The text presented here is the result of interactions with students and field veterinarians regarding stray dog welfare and public health since 2000. I am especially thankful for the collaboration of Flávia de Mello Wolff and Gisele Sprea, the veterinarians in charge of the zoonotic disease control centres of Araucária and Campo Largo, respectively, and for the collaboration of Janaina Hammerschmidt, an associated veterinarian at the Animal Welfare Laboratory, Federal University of Paraná. Comments on the initial manuscript by Clare Palmer, Mike Appleby and Dan Weary were essential for the final text format.

References

Amaku, M., Dias, R.A. and Ferreira, F. (2009) Dinâmica populacional canina: potenciais efeitos de campanhas de esterilização. *Revista Panamericana de Salud Pública* 25(4), 300–304.

Appleby, M.C. and Huertas, S.M. (2011) International issues. In: Appleby, M.C., Mench, J.A., Olsson, I.A.S. and Hughes, B.O. (eds) *Animal Welfare*, 2nd edn. CAB International, Wallingford, UK, pp. 310–312.

Beaglehole, R. and Bonita, R. (2008) Global public health: a scorecard. *Lancet* 372, 1988–1996.

Beck, A.M. (2002) *The Ecology of Stray Dogs: A Study of Free-Ranging Urban Animals*. First NotaBell Edition. Purdue University Press, West Lafayette, Indiana.

Belotto, A., Leanes, L.F., Schneider, M.C., Tamayob, H. and Correa, E. (2005) Overview of rabies in the Americas. *Virus Research* 111, 5–12.

Biondo, A.W., Cunha, G.R., Silva, M.A.G., Fuji, K.Y., Utime, R.A. and Molento, C.F.M. (2007) Carrocinha Não Resolve. *Revista do Conselho Regional de Medicina Veterinária do Paraná* 25, 20–21.

Brasil (2008) Lei No 12594 de 02 de janeiro de 2008 de Curitiba (http://www.jusbrasil.com.br/legislacao/329857/lei-12594-08-curitiba-pr, accessed 16 January 2013).

Broom, D.M. and Fraser, A.F. (2007) *Domestic Animal Behaviour and Welfare*, 4th edn. CAB International, Wallingford, UK.

Bucher, E.R. and Ranvaud, R.D. (2006) Eared dove outbreaks in South America: patterns and characteristics. *Acta Zoologica Sinica* 52(Supplement), 564–567.

Cambridge Declaration on Consciousness, The (2012) (http://fcmconference.org/img/CambridgeDeclarationOnConsciousness.pdf, accessed 16 January 2013).

Coppinger, R. and Coppinger, L. (1998) Differences in the behavior of dog breeds. In: Grandin, T. (ed.) *Genetics and the Behavior of Domestic Animals*. Academic Press, San Diego, California, pp. 167–202.

Damasco, R.T., Biondo, A.W., Koblitz, E., Fagundes, C.L., Staudacher, C., Plahinsce, C.R.S., *et al*. (2005) Controle populacional de cães na Vila Osternack, município de Curitiba, PR. *Livro de resumos do 13° EVINCI*. UFPR, Paraná, Brazil.

Davis, G.R. (2011) One health. In: Davis, G.R. (ed.) *Animals, Diseases, and Human Health: Shaping Our Lives Now and in the Future*. ABC-CLIO, Santa Barbara, California, pp. 219–238.

Dias, R.A., Garcia, R.C., Silva, D.F., Amaku, M., Ferreira-Neto, J.S. and Ferreira, F. (2004) Estimativa de populações canina e felina domiciliadas em zona urbana do Estado de São Paulo. *Revista Saúde Pública* 38(4), 565–570.

Duncan, I.J. and Petherick, J.C. (1991) The implications of cognitive processes for animal welfare. *Journal of Animal Science* 69, 5017–5022.

Gilbert, N. (2011) West Africans at risk from bat epidemics. *Nature News* (http://www.nature.com/news/2011/110922/full/news.2011.545.html, accessed 16 January 2013).

Graham, J.P., Leibler J.H., Price, L.B., Otte, J.M., Pfeiffer, D.U., Tiensin, T., *et al*. (2008) The animal–human interface and infectious disease in industrial food animal production: rethinking biosecurity and biocontainment. *Public Health Reports* 123(3), 282–299.

Hiby, L.R., Reece, J.F., Wright, R., Jaisinghani, R., Singh, B. and Hiby, E. (2011) A mark–resight survey method to estimate the roaming dog population in three cities in Rajasthan, India. *BMC Veterinary Research* 7(46), 1–9.

IBGE (Instituto Brasileiro de Geografia e Estatística) (2010) (http://www.ibge.gov.br/, accessed 14 April 2012).

Lamielle, G.A. (2010) Global Health Vet – A One Health approach to public health. Available at http://globalhealthvet.com/2010/08/21/about-one-health/ (Accessed 27 December 2013).

Landelijke Werkgroep Duivenoverlast (2006) (http://www.duivenoverlast.nl/popcycle.htm, accessed 16 January 2013).

McNeill, W.H. (1998) *Plagues and Peoples*. Bantam Doubleday Dell Publishing Group, Inc, New York.

Matos, M.R., Alves, M.C.G.P., Reichmann, M.L.A.B. and Dominguez, M.H.S. (2002) Técnica Pasteur São Paulo para dimensionamento de população canina. *Cadernos de Saúde Pública* 18(5), 1423–1428.

Miklósi, Á. (2007) *Dog Behaviour, Evolution, and Cognition*. Oxford University Press, Oxford, UK.

Mills, D.S. (2012) One health – one welfare: psychological and physical well-being. Proceedings of the WSAVA/FECAVA/BSAVA Congress 2012, Birmingham, UK. British Small Animal Veterinary Association, Gloucester, UK, pp. 237–238.

Molento, C.F.M. (2004) Vasectomising stray dogs. *Veterinary Record* 155(20), 648.

Molento, C.F.M., Lago, E. and Bond, G.B. (2007) Population control of dogs and cats in ten rural villages in the State of Paraná, Brazil: mid-term results. *Archives of Veterinary Science* 12(3), 43–50.

Morales, M.A., Varas, C. and Ibarra, L. (2009) Demographic characterization of the dog population in Viña del Mar, Chile. *Archivos de Medicina Veterinaria* 41, 89–95.

Naveda, L.A.B., Moreira, E.C., Viana, F.C., Silva, J.A. and Pereira, P.L. (2002) Avaliação do risco de Leishmaniose visceral nas comunidades de Glaura e Soares, Ouro Preto, MG. *Anais da XI Semana de Iniciaçào Científica*, Escola de Veterinária da UFMG, Minas Gerais, Brazil.

Nunes, C.M., Martines, D.A., Fikaris, S. and Queiroz, L.H. (1997) Evaluation of the dog population in an urban area of southeastern Brazil. *Revista Saúde Pública* 31(3), 308–309.

One Health Initiative (2012) (http://www.onehealthinitiative.com, accessed 6 April 2012).

Rabinowitz, P.M., Scotch, M.L. and Conti, L.A. (2010) Animals as sentinels: using comparative medicine to move beyond the laboratory. *ILAR Journal* 51(3), 262–267.

Scarlett, J. (2004) Pet population dynamics and animal shelter issues. In: Miller, L. and Zawistowski, S. (eds) *Shelter Medicine for Veterinarians and Staff*. Blackwell Publishing, Ames, Iowa, pp. 11–23.

Schneider, M.C., Almeida, G.A., Souza, L.M., Morares, N.B. and Diaz, R.C. (1996) Rabies control in Brazil from 1980 to 1990. *Revista Saúde Pública* 30(2), 196–203.

Serpell, J. (1995) *The Domestic Dog: Its Evolution, Behaviour and Interactions with People*. Cambridge University Press, Cambridge, UK.

Soto, F.R.M. (2003) Dinâmica populacional canina no Município de Ibiúna-SP: estudo restropectivo de 1998 a 2002 referente a animais recolhidos, eutanasiados e adotados. MSc thesis, University of São Paulo, Brazil.

Stafford, K. (2006) *The Welfare of Dogs*. Springer, Dordrecht, the Netherlands.

UNDP (2012) United Nations Development Programme, Human Development Reports (http://hdr.undp.org/en/countries/, accessed 8 April 2012).

WHO (2005) World Health Organization Expert Consultation on Rabies. *WHO Technical Report Series*, 931, First Report. WHO, Geneva, Switzerland.

WHO (2008) World Health Statistics 2008 (http://www.who.int/whosis/whostat/2008/en/index.html, accessed 21 January 2013).

WHO (2012a) Definition of Health (https://apps.who.int/aboutwho/en/definition.html, accessed 9 April 2012).

WHO (2012b) WHO Statistical Information System (http://apps.who.int/whosis/database/mort/table1.cfm, accessed 7 April 2012).

Zawistowski, S. and Morris, J. (2004) The evolving animal shelter. In: Miller, L. and Zawistowski, S. (eds) *Shelter Medicine for Veterinarians and Staff*. Blackwell Publishing, Ames, Iowa, pp. 3–9.

8

Balancing the Need for Conservation and the Welfare of Individual Animals

Ngaio J. Beausoleil*
Massey University, New Zealand

8.1 Abstract

Human activities and climate change have contributed to a dramatic decline in populations and species, and conservation activities are required to slow this decline. Conservation of nature is considered worthwhile by many, but for different reasons. This means that ideas about our moral obligations to protect nature, including our obligations to individual wild animals, vary. Because of this, no simple environmental ethic is likely to be adequate to guide practical decision making in conservation, particularly in situations where the protection of ecological wholes (e.g. species) impacts negatively on individual animals. Here, a practical 'ethical' approach is suggested that accommodates both the desire to conserve nature and concerns about the welfare of individual wild animals. According to this approach, our main obligation is to those sentient wild animals in whose lives we have interfered. In undertaking conservation activities that may harm individual wild animals, we are obliged to maximize the benefits of those activities and minimize any negative welfare impacts. This can be done by evaluating the relative impacts of various existing methods, choosing the most humane method, applying it in the best possible way and continuing to research more humane alternatives. This approach is illustrated by the case of the lethal control of possums in New Zealand using toxic agents. The general advantages and limitations of this 'compassionate' approach to conservation are discussed. With the continuing 'shrinking of the

*E-mail: N.J.Beausoleil@massey.ac.nz

wild', consideration of animal welfare will become increasingly important, not only to justify conservation activities but also for achieving conservation goals.

8.2 Introduction

The earth is currently experiencing a dramatic species extinction event which is largely attributed to human activity and climate change (Fa *et al.*, 2011; Harrop, 2011). The rate of known extinctions in the past 100 years is estimated to be 50–1000 times greater than the long-term average rate calculated from the fossil record, and the current rate is projected to increase tenfold in the next half century (Millennium Ecosystem Assessment, 2005).

Most agree that without conservation efforts (i.e. human interventions aimed at reversing, or at least slowing, population declines) many of those species threatened by human activities will cease to exist, at least in their present ecosystems. Conservation comprises a wide variety of activities, including those aimed at preserving wild populations in their natural habitat (*in situ* strategies), as well as complementary strategies which involve removing individuals from the wild for 'safe keeping' in captivity (*ex situ* strategies) (Table 8.1). The conservation of nature is considered worthwhile by many; however, the values people assign to nature, its components and processes vary widely. This means that ideas about our moral obligations to protect nature, including our obligations to individual wild animals, also vary.

While there is growing acknowledgement that anthropogenic activities such as agriculture and forestry impact negatively on the welfare of wild animals (Fraser, 2010a; Fraser and MacRae, 2011), the potential for conservation activities such as those described in Table 8.1 to compromise the welfare of individual wild animals has been recognized less often (however, see Norton, 1995; Sandøe and Christiansen, 2008). When conserving ecological units such as species or ecosystems comes at a cost to individual wild animals contained therein, how should we act?

This chapter will discuss one practical approach to defensible conservation decision making when such conflicts arise. It has been suggested that no simple existing environmental ethic is sufficient to provide practical guidance for conservation and that a new ethic incorporating various aspects of existing ethical theories is needed (Marks, 1999). One such hybrid ethic has been applied in the field of vertebrate pest control for at least the last decade (Littin, 2010). This 'compassionate' approach to conservation may be more widely accepted than traditional environmental ethics, as it acknowledges the potential for such activities to impact negatively on individual wild animals and aims to 'do right' by minimizing such impacts while still addressing concerns about other aspects of nature, for example the continued existence of species.

Table 8.1. Conservation-related activities. *In situ* strategies are those that occur with the animals maintained in part or all of their natural range in the wild. *Ex situ* strategies remove wild animals from their native environment and maintain them elsewhere. (From IUCN, 1998; Bellingham *et al.*, 2010; Swaisgood, 2010; Fa *et al.*, 2011.)

	General description	Conservation aims
***In situ* strategies:**		
Habitat and ecosystem protection and restoration	Legislative and physical protection and/or restoration of geographical areas that contain populations, species and ecosystems to be preserved. May involve exclusion of some or all human activities from these areas.	• Maintain numbers of animals/plants and their genetic diversity • Maintain natural relationships among species and with their physical environment (ecosystems, natural processes) • Reduce habitat destruction, degradation and fragmentation • Avoid alienation of animals from certain parts of their habitat • Reduce human–animal conflict
Intensive management of wild populations *in situ* (conservation-dependent populations)	Wild animals remain within their native range but are 'managed' intensively using one or a variety of other *in situ* procedures including: pathogen, predator or pest control; food supplementation; assisted reproduction; or physical protection, e.g. fencing.	• Maintain wild populations of species that are unable to survive in their natural range without human intervention • Avoid changes in behaviour associated with captive life

Translocation	Human-mediated movement of living organisms from one area, with release in another, intentionally and for conservation purposes. Includes movement of wild individuals within their native range and reintroduction of wild or captive-bred individuals back into their historical or current native range.	• Supplement numbers or genetic diversity or alter demography within particular wild populations (reinforcement) • Re-establish wild populations in locations from which they have been extirpated (reintroduction) • Remove individuals from locations that are no longer safe/suitable • Relieve pressure on local ecosystem by moving individuals to another location within their range
Conservation introduction ('assisted colonization' and 'ecological replacement')	Translocation of individuals of species to locations not within their traditional range. Assisted colonization refers to translocation to non-indigenous sites that are now suitable, for example, because of climate change. Ecological replacement refers to translocation to non-indigenous sites to perform a specific ecological function that has been lost through extinction (usually involves subspecies or close relative of extinct species).	• Preserve populations or species that are unlikely to survive in their current locations and when no better alternative is available • Fill 'ecological niches' left vacant by the loss of endemic species to restore ecosystems
Pest control	Lethal and non-lethal (e.g. exclusion) methods of controlling introduced populations of animals or plants considered to be pests.	• Reduce predation, competition, habitat destruction or other detrimental effects on endemic populations of plants, animals or ecosystems

(*Continued*)

Table 8.1. – (*Continued*)

	General description	Conservation aims
Culling over-abundant native species	Lethal or non-lethal (e.g. exclusion, translocation) control of over-abundant native populations that are having detrimental effects on other elements of their ecosystems. Includes prey species that may be overgrazing threatened plant species because of loss of their natural predators, or mid-level predators that have become abundant because of the loss of top-level predators or because of the abundance of introduced prey species.	• Maintain interspecific relationships to maintain ecosystems
Identification marking	Application or creation of marks to allow identification of groups or individual animals. Includes tracking/monitoring devices.	• Allow identification of groups or individual free-living animals for monitoring/tracking or research purposes
Field research	Various activities relating to the development of knowledge about free-living wild animals in their native habitat, as well as ecosystems and natural processes.	• Collect demographic data, for example, to establish population status and changes in demography, identify/prioritize habitats/ecosystems for protection • Improve understanding of the effects of human activities on wild populations • Develop and improve practical methods and tools for conserving and managing wildlife

***Ex situ* strategies:**		
Captive breeding for reintroduction	Maintenance and management of only breeding individuals in captivity for the express purpose of releasing offspring into the wild (offspring kept in captivity only for short term).	• Maximize breeding success to increase wild population numbers • Enhance survival of offspring by managing vulnerable life history phases, e.g. fledging
Captive maintenance of species	Keeping both breeding and non-breeding individuals in zoos, aquaria, wildlife parks and research facilities indefinitely.	• Maintain populations for emergency restocking of threatened wild populations or for future reintroductions of extirpated species into the wild once habitats stabilize (at some undefined point in the future) • Maintain genetic diversity in global populations • Education – increase public awareness/knowledge • Raise funds for *in situ* conservation work
Wildlife rehabilitation	Capture and treatment of injured, sick or displaced wild individuals. These individuals may be released back into the wild after treatment or maintained in captivity indefinitely.	• Maintain numbers of potential breeding individuals, and therefore genetic diversity, in wild populations
Research on captive or semi-wild animals	Various activities relating to the development of knowledge about wild species, whether applicable to life in captivity or their natural environment. Includes investigation of basic zoology (e.g. reproduction), effects of identification marking, effects of captivity on behaviour, physiology, genetic diversity, etc.	• Improve understanding of basic zoology of species relevant to protecting wild populations • Improve understanding of effects of captivity and associated procedures on wild animal health, welfare, reproduction, etc. • Improve ability to collect accurate data from free-living wild animals

To begin, I shall outline broad categories of environmental and animal welfare ethical theories relevant to the discussion of conservation decision making and will discuss briefly their limitations for guiding conservation practice. Next, I outline the underlying assumptions and general principles of the hybrid 'compassionate' approach. One popular conception of animal welfare is presented and some discussion of the evaluation of wild animal welfare, so defined, is made. I then provide an example of how this approach has been applied to improve the general acceptability of vertebrate pest control in New Zealand. Finally, I provide some discussion of the advantages and practical limits of this approach.

8.3 Nature has Value and is Worth Conserving

Nature, the natural environment, the wild, wilderness; these sorts of terms are used to describe those features, processes and characteristics of the physical world that have not been created by humans, including plants, animals and the landscape. Nature has instrumental value; that is, it is useful to humans in various ways. Instrumental values of nature include providing physical resources for current and future generations (e.g. food, fuel, medicine) and environments compatible with life (e.g. clean air, water, stable climate) (McNeely *et al.*, 1990). Some people also believe that the genetic diversity contained within nature is instrumentally valuable, as it provides the potential for adaptive change (Soule, 1985; cf. Maier, 2012).

In addition to its instrumental value, many people believe that nature or some of its components have intrinsic value and that we have moral duties to protect these values. Since the 1970s, philosophers have attempted both to apply traditional ethical systems and to devise new 'environmental ethics' to understand these moral obligations to nature and therefore guide our attempts to protect it. Ethical theories relevant to discussions of environmental issues can be categorized roughly as anthropocentric, bio/ecocentric and individualistic; these categories differ in terms of the entities or features considered to have intrinsic value, i.e. moral considerability.

8.3.1 Anthropocentric theories

Anthropocentric theories assert that only humans are intrinsically valuable and therefore morally considerable. In the context of nature conservation, anthropocentric theories suggest either that nature and its components have only instrumental value or that we have only indirect duties to protect nature; that is, if nature should be protected, it is for the sake of other humans (e.g. Passmore, 1974). Such anthropocentric theories are usually considered to be rooted in traditional Western culture, including the idea of humans' absolute dominion over nature (e.g. Genesis 1–3, The Holy Bible, King James Version).

8.3.2 Bio/ecocentric theories

Biocentric and ecocentric theories ascribe intrinsic value to components of nature (e.g. species, ecosystems or abiotic features), nature as a whole, natural processes or properties (e.g. wildness). The relative value of different elements of nature (or nature as a whole) varies depending on the specific bio/ecocentric theory. One viewpoint relevant to the present discussion is Rolston's species integrity theory, which advances species and ecosystems and their genetic integrity, rather than individuals, as the ecological units worthy of protection (Rolston, 1988).

8.3.3 Individualistic theories

Individualistic theories assign intrinsic value to individual beings or their experiences. For the purposes of this discussion, individuals considered to have intrinsic value (or whose experiences are considered valuable) are those with the capacity to experience the world in a positive or negative way, i.e. sentient animals. Both Regan's animal rights (1985) and Singer's animal liberation (1976) are individualistic theories.

According to Regan's animal rights theory, all sentient animals should be treated equally and their inherent 'rights' should not be violated for human purposes (Regan, 1985). These include the right to continuing life and the right not to be harmed. Many conservation activities would be seen by animal rights theorists as chiefly serving human interests; beyond the drive to mate and perhaps provide parental care, it is unlikely that non-human animals feel motivated to perpetuate their species or fulfil their role in ecosystem function. As such, animal rights theorists would find the protection of insentient animals, ecological wholes such as populations, species and ecosystems, abiotic features or natural processes to the detriment of individual animals morally unacceptable (Regan, 1985). One exception to this may be habitat protection (Table 8.1), which would indirectly protect the rights of individuals living therein.

In contrast to animal rights, Singer's utilitarian approach ('animal liberation') proposes that all sentient animals should have their interests considered equally; every individual's interests count and similar interests should carry similar weight (Singer, 1976). The moral acceptability of an action can then be determined by summing the satisfaction of all interests (values/benefits, e.g. pleasure) and the frustration of all interests (disvalues/costs/harm, e.g. pain); morally speaking, the optimal action is the one producing the highest ratio of benefit to cost. As such, harm to individual animals (including killing) may be permissible if the aggregated outcome has the best benefit–cost ratio.

8.3.4 Limitations to existing ethics for guiding conservation practice

In practice, environmental ethics that recognize the 'interests' of only one aspect of nature (e.g. ecological wholes or individual animals) are unlikely to provide a politically acceptable basis for conservation decision making. This is because most people reject the more radical and one-sided viewpoints, such as Rolston's species integrity theory or Regan's animal rights. Moreover, many people are intuitively dissatisfied with the results of anthropocentric evaluations, in that they fail to account for any intrinsic value of non-human animals or other components of nature. Utilitarian calculations are hampered by theoretical and practical problems, chiefly that they are impossibly complex to apply in real situations (Marks, 1999). In addition, because animal rights and utilitarian theories were developed originally to address concerns about the treatment of domestic animals (i.e. factory farming and animal experimentation), they struggle to address the very different issues associated with wild animal welfare (see the preface in Hargrove, 1992).

Hargrove (1992) notes that a practical approach to conservation is likely to be morally pluralistic and borrow from various environmental and animal welfare ethics. This sentiment is echoed by environmental philosopher Dale Jamieson (1997), who suggested that a workable environmental ethic must address concerns about both individual animals and the wider environment. As noted by Marks:

> Ecology has shown that organisms and non-living entities are inextricably linked and interdependent upon ecosystems. Sentient animals cannot be separated from these processes and their interests cannot be regarded in isolation.
>
> (Marks, 1999, p. 83)

Therefore, hybrid environmental ethics that acknowledge and attempt to balance the interests of various parties are likely to be more widely accepted and more useful for guiding defensible conservation decision making.

8.4 A Practical Approach to Improving the Welfare of Wild Animals Affected by Conservation Activities

So, in reality, conservation activities will continue because what they aim to protect is considered valuable by many people, albeit for different reasons. Conflicts between the conservation of ecological wholes and the interests of individual wild animals will occur more frequently, as both the need for human intervention to slow the rate of species extinctions (Fa *et al.*, 2011; Harrop, 2011) and acknowledgement of the potential for associated negative welfare impacts grow (Littin, 2010; Beausoleil *et al.*, 2012). The latter is reflected in increased public concern about individual wild animal welfare (e.g. Mellor *et al.*, 2004; Beausoleil and Mellor, 2007; McMahon *et al.*, 2012) and in the developing 'compassionate conservation' movement, which advocates the consideration of individual animal welfare in conservation science and management to improve outcomes for both (Born Free Foundation and Wildlife Conservation Research Unit, n.d.).

Here, a hybrid 'ethical' approach is suggested; this is not a rigorous ethical theory but rather a set of recommendations aimed at improving the general acceptability of conservation activities that have the potential to impact negatively on individual wild animals. It is based on the utilitarian ethic in that actions are not classified as right or wrong *per se*, but rather as better or worse, and incremental improvements in the welfare of wild animals affected by conservation practices are considered better than no improvements at all (Sandøe *et al.*, 1997; Littin *et al.*, 2004). This approach is consistent with animal welfare science, as it is most commonly applied to domestic animals and with the main principles of compassionate conservation.

The approach rests on several assumptions. First, it is assumed that people value some or all features of nature, because of either their instrumental or intrinsic value, or both. As such, conservation activities are undertaken to protect these valued features. Second, it is assumed that many people accept the use of animals for human purposes but are concerned about the welfare of individual wild animals for their own sake, not just because they may contribute to the conservation of nature. In other words, we assume that some individual wild animals are morally considerable (Sandøe *et al.*, 1997).

8.4.1 To which animals do we have a moral obligation?

As in animal rights and utilitarian theories, the criterion for moral considerability of wild animals is taken here to be the ability to experience mental states relevant to welfare, i.e. sentience. According to one current conception, animal welfare is conceived of, and evaluated, both at the level of the individual animal in terms of the intensity, quality and duration of its experiences and at the population level in terms of the intensity/quality/duration of individual experiences and the number of individuals affected (Gregory, 2004).

At the individual level, welfare is a state within an animal and relates to what the animal itself experiences (Mellor *et al.*, 2009; Mellor, 2012). According to this concept, welfare is considered to be the integrated balance of all sensory inputs to the animal's brain that are processed cognitively and experienced as emotions, feelings or affective/mental states. These inputs are interpreted within a framework relevant to the species and individual animal, and the integrated outcomes of this process are reflected in the individual's welfare state, which can range from very bad to very good (Mellor *et al.*, 2009; Mellor, 2012). Note that there are other approaches to the concept of animal welfare (e.g. Duncan and Fraser, 1997).

For the purposes of this discussion, all adult vertebrates are considered to meet the requisite criteria to experience at least some mental states relevant to welfare when they are conscious (Kirkwood, 2006). According to this criterion alone, we would have an obligation to consider threats to the welfare of all wild vertebrates, even those harms inflicted by predators or due to natural processes such as disease or starvation. However, making active efforts to prevent such harm is obviously impractical and, according to some, would reduce the intrinsic value of wild animals: their wildness (Norton, 1995).

So, for both practical and perhaps ethical reasons, we may have a greater moral obligation to protect the welfare of those wild vertebrate animals in whose lives we have interfered, compared to those living naturally. This is consistent with Norton's suggestion that our moral obligation increases as we act in ways that reduce the 'wildness' of wild animals (Norton, 1995). In accordance with this idea, we have more responsibility to consider and safeguard the welfare of (i.e. minimize harm to) wild animals that have been manipulated for the purposes of conservation than we do for wild animals living undisturbed. This obligation would be the same, whether the animal is considered a pest or belongs to a highly 'valued' endangered native species; these are relative classifications only. For example, the brushtail possum is considered an insidious pest in New Zealand, where it was introduced, but is a valued species in its native Australia (Warburton and Choquenot, 1999).

8.4.2 What are our obligations to these wild animals and how can we fulfil them?

So, by interfering in the lives of wild animals for the purpose of conservation, we accept a moral obligation to consider their welfare. According to a moderate utilitarian position, we can improve the moral acceptability of conservation actions by maximizing the predicted benefits of the activity and minimizing the potential costs (Marks, 1999; Littin *et al.*, 2004; Littin, 2010). The benefits of conservation activities may be accrued by the animals targeted by the activity, other animals in the environment and humans. Humans can benefit economically, aesthetically, emotionally, culturally or in terms of resource provision, both for current and future generations.

The first step to maximizing the benefits of conservation activities is the delineation of specific and measurable goals against which to measure actual conservation outcomes. These goals must be achievable; that is, the existing methods must be effective in meeting the specified goals. Once the procedure has been undertaken, the outcomes must be evaluated to determine whether the goals have been achieved and, if not, how the methods can be improved (Littin *et al.*, 2004; Clayton and Cowan, 2010).

For the purposes of this discussion, the costs or harm of conservation activities are conceived in terms of negative impacts on the welfare of individual sentient animals. As with benefits, harm can be experienced by the target animals, non-target animals and humans. Technically, costs to humans would include emotional, economic, safety and other costs; however, for the sake of clarity, these will be ignored in this discussion.

The first step to minimizing harm is to evaluate the welfare impacts associated with existing conservation methods. There are various ways of doing so, and some of the benefits and limitations of different methods for welfare assessment are discussed by Rushen and de Passillé (Chapter 10, this volume) and others (Green and Mellor, 2011; Beausoleil *et al.*, 2012; Collins, 2012). Such evaluations allow the ranking of methods in terms of their relative welfare impacts; the most humane method, i.e. the one with the fewest welfare impacts, that will achieve the conservation objectives in a given situation should be employed (Marks, 1999) (however, see discussion). They also provide a baseline for developing and assessing more humane methods (Littin *et al.*, 2004) (see below). The chosen method must then be applied in the best possible way.

8.5 Evaluation of Wild Animal Welfare

Animal welfare science traditionally has concerned itself with domestic animals and, only more recently, with captive individuals of wild species (Marks, 1999; Fraser and MacRae, 2011; Fa *et al.*, 2011). However, increasingly, animal welfare scientists are turning their attention and research methods to free-living wild animals (Mellor *et al.*, 2004; Fraser, 2010b; Littin, 2010; Swaisgood, 2010). A legacy of animal welfare science's traditional focus on domestic animals is the use of zero-suffering as the reference point against which welfare evaluations are based. Zero-suffering, which is the ultimate aim of many welfare programmes for domestic animals and captive wild animals, is considered inappropriate and impractical as a reference point for wild animal welfare (Warburton and Choquenot, 1999). Instead, these authors suggest that welfare impacts associated with conservation activities, such as vertebrate pest control, should be judged against the experiences of wild animals living undisturbed.

However, that approach could be used to support almost any treatment of wild animals, insofar as all wild animals will die eventually and death in the wild usually occurs by disease, starvation or predation (Warburton and Choquenot, 1999), all of which are likely to be preceded by negative experiences such as extreme pain, sickness or fear. There is also the problem of comparing qualitatively different experiences, e.g. those preceding 'natural' death compared with those associated with conservation activities.

Rather than attempting to judge conservation-related welfare impacts against the experiences of individuals living undisturbed, the focus of the current approach is simply to evaluate the relative impacts associated with various methods available to achieve a particular conservation goal. In doing so, the best (i.e. most humane) method can be chosen and continuing efforts made to reduce any remaining negative impacts. However, the difficulty of comparing different negative experiences remains; this is considered a problem with many, if not all, current approaches to evaluating animal welfare (see Rushen and de Passillé, Chapter 10, this volume).

Evaluating animal welfare is comparatively easy when the animals are under human control, i.e. domesticated or held in captivity. However, any assessment of wild animal welfare is limited by the inherent difficulties with collecting the necessary data from free-living individuals (Baker and Johanos, 2002; Linklater and Gedir, 2011). In addition, the focus of conservation scientists on populations and species, rather than individuals, means that most available data reflect fitness parameters, e.g. survival to breeding age, reproductive output. Such measures can be useful as they may indicate functional disruptions that also compromise welfare. However, the relationship between fitness parameters and the associated mental states that relate most directly to welfare (as defined here) will vary. For example, while reproductive failure may be indicative of poor general condition, the failure to conceive or hatch eggs *per se* may not be associated with negative emotional experiences, as it might be in humans. In contrast, it is possible for an individual to survive to breeding age but experience very poor welfare. Therefore, care must be exercised when using fitness proxies to assess wild animal welfare.

In the case study below, a framework based on the 'five domains of potential welfare compromise' was used to estimate harm. This framework provides a means of specifically considering physical or functional impacts on the animal and their contribution to the affective experiences, mental states or feelings that ultimately determine its welfare according to the concept outlined above (Mellor and Reid, 1994; Mellor *et al.*, 2009). This method can be used to estimate the relative intensity, quality and duration of individual wild animals' experiences.

8.6 Lethal Possum Control Using Toxins in New Zealand

8.6.1 Background

New Zealand (NZ) has a unique terrestrial vertebrate fauna with a very high level of endemism, i.e. species found nowhere else in the world. Accordingly, local extirpation usually represents global extinction. One of the important consequences of NZ's long geographical isolation is the failure of endemic fauna, particularly birds, to develop antipredator behaviour against mammals, making them exceptionally vulnerable to introduced predators. In fact, mammalian predation is considered the primary cause of population decline and extinction of NZ vertebrates (Innes *et al.*, 2010), and pest control has been recognized as a priority for NZ conservation for at least 100 years (Bellingham *et al.*, 2010).

The brushtail possum (*Trichosurus vulpecula*; Fig. 8.1) was introduced to NZ from Australia in 1837 to establish a fur trade. Populations quickly expanded and now occupy more than 95% of NZ's land area (Lawton *et al.*, 2010). Possums pose conservation and economic threats, by destroying habitat required by many native species and by eating their eggs and chicks, as well as by acting as a wildlife reservoir for important agricultural diseases such as bovine tuberculosis. For these reasons, various NZ agencies now control possums over approximately 8 million ha of the country (Landcare Research NZ Ltd, 2008).

Control of possums using lethal toxins was chosen to illustrate how 'ethical' conservation can be conducted for several reasons. The uniqueness of the fauna, along with the relatively recent arrival of humans and a clearly correlated acceleration in extinction rates, has created a strong 'culture of conservation' in NZ. As a result, NZ invests heavily in, and has strong public support for, conservation activities, providing a wealth of information with which to evaluate welfare impacts. With regard to possum control, a significant body of published information relevant to welfare assessment is available, and recent

Fig. 8.1. Brushtail possum (*Trichosurus vulpecula*). Photo copyright Bruce Warburton.

welfare evaluations have been comparatively comprehensive, yielding results well supported by current scientific evidence (Sharp and Saunders, 2008; Beausoleil *et al.*, 2010, 2012).

8.6.2 Benefits of lethal possum control

The explicit conservation goals of possum control in NZ are to restore the habitat and populations of native species. Success is measured in terms of operational outcomes (e.g. residual possum numbers after control) and performance outcomes (e.g. abundance of birds and improvement in forest canopy health) (Clayton and Cowan, 2010). Mammalian pest control in NZ is considered successful if residual numbers are sufficiently low that improvements in native species populations and/or habitat health can be measured (Lawton *et al.*, 2010). These goals are specific and measurable, but are they achievable, i.e. are the current methods effective in achieving these goals?

In theory, possums can be controlled using lethal or non-lethal methods. Lethal methods include poisoning (toxins), trapping and shooting, while non-lethal methods include exclusion fences and immunocontraception. In reality, there are currently no practical alternatives to lethal methods for controlling possums, and toxins are the most efficacious way to reduce numbers sufficiently to meet the specified goals (Warburton and Choquenot, 1999; Twigg and Parker, 2010).

The three most commonly used toxins for the lethal control of possums are cyanide, sodium fluoroacetate ('1080') and brodifacoum, an anticoagulant poison (Eason *et al.*, 2010). Possum control using these methods has been successful in meeting conservation targets in certain locations (Parkes *et al.*, 1997). The most dramatic improvements are reported when possums (and other mammals) are eradicated completely from small islands or fenced mainland sanctuaries with no subsequent immigration (Reddiex *et al.*, 2007). However, for open populations, benefits are temporary as possums quickly reinvade (Lawton *et al.*, 2010; see also discussions of feral population control by Molento, Chapter 7, and Palmer, Chapter 9, this volume). Therefore, in many locations, maintaining initial benefits depends on continuing control efforts (Parkes *et al.*, 1997) and continuing potential for welfare impacts on possums (see below).

Lethal possum control benefits native bird (and probably other vertebrate) species. Without the effective control of multiple mammalian pest species, including possums, many native NZ bird species rapidly would become extinct. This has become apparent from historical temporal associations between pest introductions and declines in bird populations and from examining the effects of recent pest control operations on current bird populations (Innes *et al.*, 2010). While going extinct per se is not a welfare issue for individual animals, the interdependence of species, ecosystems and natural processes means that individuals are increasingly likely to encounter situations that compromise their welfare before extinction occurs. For example, the consequences of declining population sizes (e.g. inbreeding depression) and the proximate causes of extinction (e.g. disease epidemics, inadequate food supplies) will compromise welfare. Therefore, slowing population declines by eradicating possums will benefit individual native birds.

As well as native birds, domestic cattle may benefit, as reducing possum populations is believed to reduce the transmission of bovine tuberculosis (Landcare Research NZ Ltd, 2008) and associated negative welfare impacts such as sickness. Humans may benefit economically from reduced prevalence of bovine TB, and in terms of tourism when the habitats and populations of native animal species are maintained or restored. Humans may also benefit emotionally from the knowledge that the intrinsic value of the environment is being protected, and indigenous Maori may benefit culturally from the continued existence of some native species.

8.6.3 Costs of lethal possum control

The costs of lethal control are evaluated most easily for the target animals themselves. A systematic evaluation demonstrates that there are welfare impacts associated with all toxins currently used to control possums in NZ (Beausoleil *et al.*, 2010, 2012). The impacts range from moderate breathlessness, lasting for minutes with cyanide, to severe or extreme pain, lethargy and sickness, lasting for days to weeks with the anticoagulant toxins. 1080 had intermediate effects, although these are currently not well understood. Importantly, even cyanide, judged to have the lowest welfare impacts of those assessed (Littin *et al.*, 2009; Eason *et al.*, 2010), causes some negative affective experiences before loss of consciousness and death (Beausoleil *et al.*, 2010). In practice, this is likely to be true for all lethal pest control methods (Littin, 2010).

In addition to these impacts on possums consuming poison baits, there may be inadvertent welfare costs to other animals. These include starvation of dependents (i.e. young in pouch), poisoning of other pest (e.g. deer) or domestic (e.g. dogs and livestock) species, either through ingestion of bait or of poisoned carcasses, and the lethal and sublethal poisoning of native birds (Littin *et al.*, 2004). Non-target species may have different susceptibility to the toxic agents, with their welfare and fitness impacted to a greater or lesser degree (Eason *et al.*, 2010). In addition, it has been suggested that removing some or all pests may have unexpected effects on the ecosystems into which they have become integrated, some of which may impact negatively on native animals (Innes *et al.*, 2010). For example, removal of one pest species may allow another to take over, which may have even greater impacts on various native species (Littin *et al.*, 2004).

8.6.4 How can benefits be maximized and harm minimized to improve acceptability?

Overall, the goals and methods of lethal pest control using toxins appear appropriate to maximize its benefits, although maintaining benefits in open populations depends on continuing control efforts. To maximize the benefits of pest control programmes, goals must reflect the desired reduction in the ecological impacts of the pest, rather than simply prescribing an arbitrary 'kill quota' (Marks, 1999). Strategic sustained management aims to reduce a pest population to an initial low abundance and suppress population growth using regular maintenance control. The overall goal of this strategy is to maximize the conservation benefits while killing the minimum number of individual pests over time (Williams *et al.*, 1995), which is particularly important for open populations.

All of the methods currently used to poison possums are associated with significant welfare impacts before loss of consciousness and death, but cyanide has the least severe impacts. Therefore, to maximize the acceptability of possum poisoning, cyanide should be used wherever possible and the use of anticoagulant poisons avoided. However, in reality, use of the most humane method is not always possible. For example, aerial bait distribution is the only practical method of possum control in many remote locations in NZ; of the effective poisons, only 1080 is registered for aerial distribution on mainland NZ and must currently be used, despite its greater welfare impacts. In such cases, there may be other ways to ameliorate some of the negative impacts associated with poisoning. For example, inclusion of an analgesic was found to reduce welfare impacts on red foxes poisoned with 1080 (Marks *et al.*, 2009).

Obvious and short-term impacts on non-targets, particularly native birds, can be minimized by the way poison baits are formulated, presented and distributed. For example, accidental poisoning of native birds can be reduced by altering the colour, flavour and matrix of the bait and by the use of bait stations (Eason *et al.*, 2010). In addition, warning owners of the presence and danger of poison baits and judicious application of baits near recreational or agricultural areas will minimize impacts on the welfare of domestic animals such as dogs and livestock (Eason *et al.*, 2010). However, longer-term welfare impacts resulting from changes in ecosystem function due to the removal of possums will be difficult to assess and rectify, e.g. starvation of native species (Innes *et al.*, 2010).

In the spirit of harm minimization, scientists also have an obligation to continue to research the welfare impacts of existing toxins, many of which are still poorly understood (Beausoleil *et al.*, 2010; Beausoleil and Mellor, 2012), refine existing methods and investigate effective new methods for the lethal and non-lethal control of possums (Littin *et al.*, 2004). This research approach is consistent with the three Rs principles for harm minimization in laboratory animal science (Russell and Burch, 1959). Note that this type of research also has the potential to impact negatively on the welfare of the subjects and the same harm minimization principles should be applied.

8.7 General Discussion

Obviously, the 'compassionate' approach presented is only one possibility for guiding 'ethical' conservation practice. One advantage of this hybrid ethical approach is that it is not restricted by competing value bases associated with discrete environmental ethical theories (however, see below for one exception). Use of this approach permits conservation activities such as pest control, despite the negative impacts on individual animals, thereby acknowledging implicitly that nature has value, whatever its form, and that its protection is important to many people and/or for its own sake. This acknowledgement will make such activities generally acceptable from a bio/ecocentric perspective. In addition, the approach may be seen as consistent with anthropocentric theories in that the protection of nature can be seen as an indirect duty to other humans, e.g. imparting emotional benefits to those that care about nature in addition to safeguarding its instrumental value.

At the same time, the approach acknowledges the potential for such activities to reduce the welfare of individuals of wild species and requires that any suffering be minimized (however, see below). This will probably improve the acceptability of these activities to those concerned with wild animals' welfare, such as those following a moderate utilitarian ethic. However, this approach does not require the strict summation of predicted benefits and costs to decide whether an action increases the overall good in the world.

Some actions permitted by this approach will not be acceptable to those adhering to an animal rights theory. Implicit in this approach is the acceptance of interference with wild animals for conservation purposes, even when such interference may harm or even kill some animals. Despite attempts to minimize negative experiences before death, animal rights proponents would see actions such as lethal mammalian pest control as violating the rights of sentient animals to continue living and to be free from harm at the hands of humans. However, this position may be problematic. Some would argue that in actively introducing pest animals, we have violated the rights of native species. Refraining from controlling these pests could then be seen as further violation of the rights of native birds, as the harm done to them is due ultimately to the actions of humans. It therefore seems impossible to respect the rights of all qualifying individuals in such situations, making practical decision making in conservation impossible.

In reality, applying the approach will not be as easy as has been implied. For example, the most humane method is not always the most efficacious, and welfare impacts have to be balanced against conservation goals and other factors, including human safety (Warburton, 1998). In the example of lethal possum control, conservation goals appear to prevail over concerns for animal welfare: when the specified goal cannot be achieved using the most humane method (i.e. cyanide), a less humane method is used (e.g. 1080) rather than abandoning the activity or revising the goal. This reflects the fact that, like many of the traditional environmental ethics, hybrid approaches still prioritize the interests of some parties over others. In practice, the principle of harm minimization is likely to be applied only as it is consistent with achieving conservation and other goals that satisfy human interests.

However, recall that the 'compassionate' approach presented here is based on the notion that any improvement in animal welfare is better than none. For possum control, 1080, although not the most humane of the currently used poisons, is the most effective for reducing possum populations in remote locations. Although less humane than cyanide, 1080 is more humane than the anticoagulant poisons. Therefore, it should be used in preference to the anticoagulant poisons, but cyanide should be used wherever possible.

In addition, the perennial problem of comparing qualitatively different experiences persists. In the case of possum poisoning, ranking the various agents was relatively easy as the durations of effects were very different (minutes for cyanide versus weeks for anticoagulants). For other conservation activities, ranking may be more difficult.

Nevertheless, this practical 'ethical' approach provides a way to satisfy, at least partially, the interests of both humans who value the environment and those concerned about the welfare of individual wild animals. Explicit consideration of animal welfare is critical to drive the development of more welfare-friendly conservation methods, which may also improve conservation outcomes such as survival and reproduction in vulnerable species. With the continuing 'shrinking of the wild', the objectives and activities of wildlife conservation and animal welfare increasingly will overlap.

8.8 Conclusions

- The earth is experiencing a dramatic extinction event which is attributed to human activity and climate change.
- Without targeted efforts to halt this decline, many of those species threatened by human activities will cease to exist, at least in their present ecosystems.
- Conservation is considered necessary by many people because nature and its components are valued, although for different reasons. However, there are conflicts between the conservation of ecological wholes and the welfare of individual animals.
- Ethical systems relevant to conservation can be categorized broadly as anthropocentric, bio/ecocentric and individualistic, emphasizing human

interests, the intrinsic value(s) of nature and the interests of non-human animals, respectively. No one ethic is likely to be an acceptable basis for practical decision making in conservation.
- A hybrid 'ethical' approach is suggested that accommodates both the desire to conserve nature and concerns about the welfare of individual wild animals.
- According to this approach, our main obligation is to those sentient wild animals in whose lives we have interfered.
- This obligation can be fulfilled by maximizing the benefits of conservation activities and minimizing any harm, including negative impacts on the welfare of individual wild animals. In practice, this means evaluating the relative impacts of various methods and choosing the most humane. Lethal control of possums in New Zealand using toxic agents is used to illustrate this point.
- The main advantage of this 'compassionate' approach is that it acknowledges both the value of nature and the welfare of individual wild animals. However, as with other environmental ethics, the outcomes of this hybrid approach will not be acceptable to everyone.
- With the continuing 'shrinking of the wild', consideration of animal welfare will become increasingly important, not only to justify conservation activities but also for achieving conservation goals.

Questions for discussion

1. Imagine that a native bird species that you care about is threatened by introduced carnivores. To save the birds, it seems necessary to use poison to control the predators, even in periods when their young may be exposed. Would you endorse this course of action?
2. How can the welfare of a few rare, native, sentient individual animals be balanced against that of numerous sentient pests when these interests are in conflict? More generally, do we have a different moral obligation to consider the welfare of wild animals classified differently by humans, e.g. pests and valued species?
3. Should seriously injured or sick individual wild animals be rehabilitated for conservation purposes or should they be euthanized? Does the answer depend on whether they are members of an endangered or a common species?
4. Do intensively managed 'wild' animals have acceptable welfare? (See Table 8.1 for descriptions of intensive management.) Is such intensive interference justified to save individuals of rare species?

Acknowledgements

Sincere thanks to David Mellor for detailed and constructive editing of this manuscript and to Peter Sandøe and Clare Palmer for critical review. Thanks also to Kate Littin and Penny Fisher for advice relevant to the welfare evaluations of possum control and to Bruce Warburton for providing the photograph.

References

Baker, J.D. and Johanos, T.C. (2002) Effects of research handling on the endangered Hawaiian monk seal. *Marine Mammal Science* 18, 500–512.

Beausoleil, N.J. and Mellor, D.J. (2007) Investigator responsibilities and animal welfare issues raised by hot branding of pinnipeds. *Australian Veterinary Journal* 85(12), 484–485.

Beausoleil, N.J. and Mellor, D.J. (2012) Complementary roles for systematic analytical evaluation and qualitative whole animal profiling in welfare assessment for Three Rs applications. Session IV–1: Indicators of Animal Welfare to Implement Refinement. In: von Aulock, S. (ed.) *Proceedings of the 8th World Congress on Alternatives and Animal Use in the Life Sciences*, Montreal, Canada, 21–25 August 2011. *ALTEX* Proceedings 1 (WC8). Springer Spektrum, Heidelberg, Germany, pp. 455–460.

Beausoleil, N.J., Fisher, P., Warburton, B. and Mellor, D.J. (2010) Vertebrate toxic agents and kill traps in mammal species. In: *How Humane Are Our Pest Control Tools? MAF Biosecurity New Zealand Technical Paper No: 2011/01*. MAF Biosecurity New Zealand, Wellington, pp. 1–102.

Beausoleil, N.J., Fisher, P., Mellor, D.J. and Warburton, B. (2012) Ranking the negative impacts of wildlife control methods may help advance the Three Rs. Theme IV–3: Wildlife Welfare and the Three Rs. In: von Aulock, S. (ed.) *Proceedings of the 8th World Congress on Alternatives and Animal Use in the Life Sciences*, Montreal, Canada, 21–25 August 2011. *ALTEX* Proceedings 1 (WC8). Springer Spektrum, Heidelberg, Germany, pp. 481–485.

Bellingham, P.J., Towns, D.R., Cameron, E.K., Davis, J.J., Wardle, D.A., Wilmshurst, J.M., *et al.* (2010) New Zealand island restoration: seabirds, predators, and the importance of history. *New Zealand Journal of Ecology* 34(1), 115–136.

Born Free Foundation and Wildlife Conservation Research Unit (n.d.) Compassionate conservation: animal welfare in conservation practice (www.compassionateconservation.org, accessed 12 July 2012).

Clayton, R. and Cowan, P. (2010) Management of animal and plant pests in New Zealand – patterns of control and monitoring by regional agencies. *Wildlife Research* 37, 360–371.

Collins, L.M. (2012) Welfare risk assessment: the benefits and common pitfalls. *Animal Welfare* 21(Supp 1), 73–79.

Duncan, I.J.H. and Fraser, D. (1997) Understanding animal welfare. In: Appleby, M.C. and Hughes, B.O. (eds) *Animal Welfare*. CAB International, Wallingford, UK, pp. 19–31.

Eason, C.T., Miller, A., Ogilvie, S. and Fairweather, A. (2010) An updated review of the toxicology and ecotoxicology of sodium fluoroacetate (1080) in relation to its use as a pest control tool in New Zealand. *New Zealand Journal of Ecology* 35, 1–20.

Fa, J.E., Funk, S.M. and O'Connell, D. (2011) *Zoo Conservation Biology*. Cambridge University Press, Cambridge, UK.

Fraser, D. (2010a) Conservation and animal welfare. *Animal Welfare* 19, 121–211.

Fraser, D. (2010b) Toward a synthesis of conservation and animal welfare science. *Animal Welfare* 19, 121–124.

Fraser, D. and MacRae, A.M. (2011) Four types of activities that affect animals: implications for animal welfare science and animal ethics philosophy. *Animal Welfare* 20, 581–590.

Green, T.C. and Mellor, D.J. (2011) Extending ideas about animal welfare assessment to include 'quality of life' and related concepts. *New Zealand Veterinary Journal* 59, 263–271.

Gregory, N.G. (2004) *Physiology and Behaviour of Animal Suffering*. Blackwell Scientific Publishing, Oxford, UK.

Hargrove, E.C. (ed.) (1992) *The Animal Rights/Environmental Ethics Debate: The Environmental Perspective*. State University of New York Press, New York.

Harrop, S. (2011) Climate change, conservation and the place for wild animal welfare in international law. *Journal of Environmental Law* 23, 441–462.

Innes, J., Kelly, D., Overton, J.M. and Gillies, C. (2010) Predation and other factors currently limiting New Zealand forest birds. *New Zealand Journal of Ecology* 34, 86–114.

IUCN (1998) *IUCN Guidelines for Reintroductions*. IUCN/SSC Reintroduction Specialist Group, Gland, Switzerland.

Jamieson, D. (1997) Animal liberation is an environmental ethic (http://www.acad.carleton.edu/curricular/ENTS/faculty/dale/dale_animal.html, accessed 15 August 2012).

Kirkwood, J. (2006) The distribution of the capacity for sentience in the animal kingdom. In: Turner, J. and D'Silva, J. (eds) *Animals, Ethics and Trade: The Challenge of Animal Sentience*. Compassion in World Farming, Earthscan, London, pp. 12–26.

Landcare Research NZ Ltd (2008) Possums – their introduction and spread (http://www.landcareresearch.co.nz/publications/infosheets/possums/introduction_spread.pdf, accessed 29 March 2012).

Lawton, C., Cowan, P., Bertolino, S., Lurz, P.W.W. and Peters, A.R. (2010) The consequences of introducing non-indigenous species: two case studies, the grey squirrel in Europe and the brushtail possum in New Zealand. *Review of Science and Technology Office International Epizooties* 29, 287–298.

Linklater, W.L. and Gedir, J.V. (2011) Distress unites animal conservation and welfare towards synthesis and collaboration. *Animal Conservation* 14, 25–27.

Littin, K.E. (2010) Animal welfare and pest control: meeting both conservation and animal welfare goals. *Animal Welfare* 19, 171–176.

Littin, K.E., Mellor, D.J., Warburton, B. and Eason, C.T. (2004) Ethical and animal welfare principles for humane vertebrate pest control: a review. *New Zealand Veterinary Journal* 52, 1–10.

Littin, K.E., Gregory, N.G., Airey, A.T., Eason, C.T. and Mellor, D.J. (2009) Behaviour and time to unconsciousness of brushtail possums (*Trichosurus vulpecula*) after a lethal or sublethal dose of 1080. *Wildlife Research* 36, 709–720.

McMahon, C.R., Harcourt, R., Bateson, P. and Hindell, M.A. (2012) Animal welfare and decision making in wildlife research. *Biological Conservation* 153, 254–256.

McNeely, J.A., Miller, K.R., Reid, W.V., Mittermeier, R.A. and Werner, T.B. (1990) *Conserving the World's Biological Diversity*. IUCN, Gland, Switzerland, Washington, DC.

Maier, D. (2012) *What's So Good About Biodiversity?A Call for Better Reasoning About Nature's Value*. International Library of Environmental, Agricultural and Food Ethics, 19. Springer, New York.

Marks, C. (1999) Ethical issues in vertebrate pest management: can we balance the welfare of individuals and ecosystems? In: Mellor, D.J. and Monamy, V. (eds) *The Use of Wildlife for Research. Proceedings of the ANZCCART Conference, 26–27 May, 1999*. Australian and New Zealand Council for the Care of Animals in Research and Teaching (ANZCCART), Dubbo, NSW, Australia, pp. 79–89.

Marks, C., Gigliotti, F. and Busana, F. (2009) Assuring that 1080 toxicosis in the red fox (*Vulpes vulpes*) is humane. *Wildlife Research* 36, 98–105.

Mellor, D.J. (2012) Animal emotions, behaviour and the promotion of positive welfare states. *New Zealand Veterinary Journal* 60, 1–8.

Mellor, D.J. and Reid, C.S.W. (1994) Concepts of animal well-being and predicting the impact of procedures on experimental animals. In: Baker, R.M., Jenkin, G. and Mellor, D.J. (eds) *Improving the Well-being of Animals in the Research Environment*. Australian and New Zealand Council for the Care of Animals in Research and Teaching, Glen Osmond, South Australia, pp. 3–18.

Mellor, D.J., Beausoleil, N.J. and Stafford, K.J. (2004) *Marking Amphibians, Reptiles and Marine Mammals: Animal Welfare, Practicalities and Public Perceptions in New Zealand*. Department of Conservation, Wellington.

Mellor, D.J., Patterson-Kane, E. and Stafford, K.J. (2009) *The Sciences of Animal Welfare*. Wiley-Blackwell Publishing, Oxford, UK.

Millennium Ecosystem Assessment (2005) *Ecosystems and Human Well-being: Synthesis*. Island Press, Washington, DC.

Norton, B.G. (1995) Caring for nature: a broader look at animal stewardship. In: Norton, B.G., Hutchins, M., Stevens, E.F. and Maple, T.L. (eds) *Ethics on the Ark: Zoos, Animal Welfare, and Wildlife Conservation*. Smithsonian Institution Press, Washington, DC, pp. 102–121.

Parkes, J., Baker, A.N. and Ericksen, K. (1997) *Possum Control by the Department of Conservation. Background, Issues and Results from 1993 to 1995*. Department of Conservation, Wellington.

Passmore, J.A. (1974) *Man's Responsibility for Nature: Ecological Problems and Western Traditions*. Charles Scribner's Sons, New York.

Reddiex, B., Fraser, W., Ferriss, S. and Parkes, J. (2007) *Animal Health Board Possum Control Operations on Public Conservation Lands: Habitats Treated and Resulting Possum Abundance. DoC Research and Development Series 277*. Department of Conservation, Wellington.

Regan, T. (1985) The case for animal rights. In: Singer, P. (ed.) *In Defense of Animals*. Basil Blackwell, Oxford, UK, pp. 13–26.

Rolston, H. (1988) *Environmental Ethics: Duties to and Values in the Natural World*. Temple University Press, Philadelphia, PA.

Russell, W.M.S. and Burch, R.L. (1959) *The Principles of Humane Experimental Technique*. Methuen and Co, London.

Sandøe, P. and Christiansen, S.B. (2008) Management and use of wild animals. In: Sandøe, P. and Christiansen, S.B. (eds) *Ethics of Animal Use*. Wiley-Blackwell, Chichester, UK, pp. 153–170.

Sandøe, P., Crisp, R. and Holtug, N. (1997) Ethics. In: Appleby, M.C. and Hughes, B.O. (eds) *Animal Welfare*. CAB International, Wallingford, UK, pp. 3–17.

Sharp, T. and Saunders, G. (2008) *A Model for Assessing the Relative Humaneness of Pest Animal Control Methods*. Australian Government Department of Agriculture, Fisheries and Forestry, Canberra, ACT.

Singer, P. (1976) All animals are equal. In: Regan, T. and Singer, P. (eds) *Animal Rights and Human Obligations*. Prentice-Hall Inc, Upper Saddle River, NJ, pp. 148–162.

Soule, M.E. (1985) What is conservation biology? *BioScience* 35, 727–734.

Swaisgood, R.R. (2010) The conservation–welfare nexus in reintroduction programmes: a role for sensory ecology. *Animal Welfare* 19, 125–137.

Twigg, L.E. and Parker, R.W. (2010) Is sodium fluoroacetate a humane poison? The influence of mode of action, physiology, effect and target specificity. *Animal Welfare* 19, 249–263.

Warburton, B. (1998) The 'humane' trap saga: a tale of competing ethical ideologies. In: Mellor, D.J., Fisher, M. and Sutherland, G. (eds) *Ethical Approaches to Animal-Based Science. Proceedings of the Conference of the Australasian and New Zealand Council for the Care of Animals in Research and Teaching (ANZCCART)*, 19–20 September 1997, Auckland, New Zealand. ANZCCART, Wellington, pp. 131–137.

Warburton, B. and Choquenot, D. (1999) Animal welfare and pest control – the context is important. In: Mellor, D.J. and Monamy, V. (eds) *The Use of Wildlife for Research. Proceedings of the ANZCCART Conference, 26–27 May, 1999*. Australian and New Zealand Council for the Care of Animals in Research and Teaching (ANZCCART), Dubbo, NSW, Australia, pp. 90–99.

Williams, K., Parer, I., Coman, B., Burley, J. and Braysher, M. (1995) *Managing Vertebrate Pests: Rabbits*. Australian Government Publishing Service, Canberra.

9

Value Conflicts in Feral Cat Management: Trap–Neuter–Return or Trap–Euthanize?

Clare Palmer*
Texas A&M University, USA

9.1 Abstract

This chapter explores the key values at stake in feral cat management, focusing on the debate over whether to use trap–neuter–return or trap–euthanize as management tools for cat populations. The chapter provides empirical background on unowned cats, sketches widely-used arguments in favour of reducing cat populations and considers how these arguments relate to important and widely held values including the value of lives, subjective experiences and species. The chapter promotes critical understanding of the diverse value positions that may be at play in debates about the treatment of ownerless cats.

9.2 Introduction

The management of ownerless cats is highly controversial. Ownerless cats can be regarded as messy pests, threats to public health, profligate hunters of already vulnerable wildlife, integrated members of novel ecosystems, intelligent and sentient independent actors or abandoned and suffering victims of human neglect. Each of these interpretations of ownerless cats prioritizes certain values and disvalues – such as avoiding suffering or species protection – over others. These different value priorities affect not only how ownerless cats are perceived but also what practices are proposed to manage them.

---

*E-mail: cpalmer@philosophy.tamu.edu

This chapter will explore the underlying values in debates about cat management. To focus a potentially broad discussion, the chapter will concentrate on trap–neuter–return (TNR) – trapping cats, neutering them and returning them to their original location – as a management policy.[1] Many animal welfare organizations and feral cat advocates, such as Alley Cat Allies (ACA) and The Humane Society of the United States (HSUS), regard TNR as the best way of controlling unowned cat populations and maintaining cat welfare, with associated benefits for public health. In contrast, wildlife conservationists argue that TNR perpetuates wildlife destruction by unowned cats and that we should therefore adopt trap–euthanize (TE) policies. Some animal rights advocates also argue for TE over TNR, if cats cannot otherwise be homed, primarily on the grounds that this is better for cats themselves.[2]

I will begin by providing some background information on unowned cats and on TNR and TE as management tools. Then, I will sketch widely used arguments in favour of reducing unowned cat populations, briefly considering public health issues (see also Molento, Chapter 7, this volume) but concentrating on cat welfare and wildlife concerns. I will consider the 'effectiveness' of both TNR and TE at addressing these concerns. This will open up discussion about the underlying values at stake. I will focus on three values in particular: animal lives, subjective animal welfare and species. Exploring these values should assist in clarifying and explaining underlying justifications for the contrasting positions adopted in this debate, and in highlighting relevant value considerations when creating policies to manage unowned cats. It should be noted, though, that the chapter is intended to be exploratory and explanatory, rather than prescriptive.

9.3 Feral Cats, Trap–Neuter–Return and Trap–Euthanize

Recent statistics estimate that there are 86.4 million owned cats in the USA and that 33% of households contain at least one cat (HSUS, 2011). The UK has about 8 million owned cats (PFMA, 2011). Alongside this 'owned' population, there is a substantial 'unowned' population (and some cats do not fall easily into either category). Some unowned cats are strays, and may be reasonably well socialized to humans, but others have never had human contact. After about 8 weeks of age, such cats are extremely difficult to socialize or adopt and are regarded as 'truly feral' (Slater, 2007; ASPCA, 2012). It is estimated that 36–41% of the total cat population in the USA is unowned (Scott *et al.*, 2002a); indeed, estimates for the total number of unowned cats in the USA alone reach as high as 70–100 million (Jessup, 2004; Mott, 2004). For simplicity, all unowned cats shall be called 'feral' here; this is not meant to deny that some may be wholly or partially socialized.

Cats can reproduce prolifically. A queen may begin reproducing at 6 months and have 3 litters a year. The average feral queen, though, has fewer litters; one study reported 1.4 litters per queen annually (Stoskopf and Nutter, 2004). Litters vary between 2 and 5 kittens. Stoskopf and Nutter (2004) found live births to average 3.5 +/– 1.2 kittens, but these litters had low survival rates: cumulative feral kitten mortality is 75% at 6 months. Despite these high mortality rates, still higher reproductive rates mean that feral cat populations can increase rapidly (as Molento, Chapter 7, this volume, suggests is also true of feral dogs). This brings us to TNR, a strategy that is intended to control feral cat populations (Fig. 9.1).

All TNR programmes trap and neuter unowned cats. Most also vaccinate them, especially against rabies; some also deworm. A minority of programmes test for feline leukaemia (FeLV) and feline AIDS (FIV) and may euthanize cats that are sick or are carriers. Other TNR programmes remove kittens and adoptable cats to home them (this is sometimes called TNR+). Almost all programmes remove the left ear tip to indicate that the cats have been neutered (Cuffe, 1983). The cats are then returned to the colony. Many TNR programmes will only return cats to original colonies if reliable sources of food, water and shelter from carers have been identified (ASPCA, 2012). In principle, trapped cats could be relocated; but, in practice, there are rarely plausible alternative locations.

TE programmes also trap cats, but unadoptable cats are euthanized. Even adoptable cats probably will ultimately be euthanized, given the high

Fig. 9.1. Reproductive activity in a non-neutered feral cat colony.

euthanasia rates in many animal shelters; in California, for instance, 81% of cats impounded in open access shelters were ultimately euthanized (Kass, 2005). Alternative strategies to control cat populations do exist, including poison bait and biological control. But these risk impacting non-target animals and are only practical on islands or in very isolated populations (Slater, 2007). So, in most cases, especially in urban/suburban contexts, the alternative to TNR is TE.

Both TNR and TE, then, have as their goal the reduction in size, or the elimination, of feral cat colonies. But why reduce them at all?

9.4 Why Reduce Feral Cat Populations?

9.4.1 Negative impacts on humans

There are two main objections to feral cats in human communities: that they are a significant nuisance and that they act as vectors of disease.

Feral cats may generate noise, mess and smell that cause disturbance or nuisance to people nearby. They do, though, also hunt other animals widely regarded as pests, in particular rodents, and they generate positive responses in some people who interact with them (Slater, 2007). More significant than causing nuisance, feral cats carry disease. They can catch, and carry, rabies, though they are not the primary vectors for the disease (Slater, 2007; see Molento, Chapter 7, this volume, on rabies). Cat faeces can contain toxoplasmosis, a disease with low incidence but dangerous to vulnerable groups (Afonso *et al.*, 2006). Feral cats can host fleas, ear mites, hookworms and roundworms, and can serve as a reservoir for Bartonella infections (including cat scratch disease), which may pass from the feral to the homed cat population, as well as to other wildlife and ultimately to humans. They may carry, or suffer from, FelV or FIV, which, while not transmissible to humans, could infect homed cats that go outdoors.

TNR colonies generate fewer of these concerns. Neutering reduces nuisance, as neutered male cats display fewer spraying, vocalizing and aggressive behaviours (Hart and Barrett, 1973). It should also reduce disease transmission, as mating fights and giving birth are key transmission routes for FIV and FeLV. If cats are wormed and vaccinated for rabies when trapped, this further reduces the likelihood of disease transmission. So, TNR colonies should be less problematic than other feral cat colonies.

9.4.2 Feral cat welfare is poor

A second argument for reducing feral cat populations is that the welfare of feral cats themselves, in terms of longevity, disease and risk of physical trauma, is so poor that population reduction is in the cats' own interests. (The meaning of 'welfare' is considered later.) Stoskopf and Nutter (2004), for instance, found a median survival time of 113 days among the 75% of feral kittens that did not survive to 6 months of age. Some of these kittens may have died suddenly, but probably many had lifelong poor welfare. One idea of TNR colonies is, of course, that such miserable kittens never come into existence. Jessup (2004), a critic of TNR, points to an American Veterinary Medical Association figure of a 2-year average lifespan for feral cats, in comparison with an average 10-year lifespan for owned cats. He cites claims by the animal rights organization, People for the Ethical Treatment of Animals (PETA), that feral cats have 'horrific fates': they 'get heart disease, leukaemia, bladder problems, ear infections, and more ... Contagious diseases such as rhinotracheitis, feline AIDS, and rabies are common in "outdoor cats", who also sustain puncture wounds, broken bones, brain damage, or loss of an eye or limb after they are attacked by other animals or hit by cars' (Jessup, 2004, p. 1378). Both Jessup (2004) and PETA (2012) maintain that this concern about poor welfare applies to all feral cats, including those in TNR colonies, though neither cites detailed studies to support their accounts of feral cats' welfare, either within or outside TNR colonies. In fact, few reviewed published studies on feral cat welfare exist.

There is some evidence about feral cat welfare in TNR colonies. Scott *et al.* (2002b), for instance, noted that feral cats were lean but not emaciated before being neutered and that, after neutering, feral cats, like owned cats, gained weight. Centoze and Levy (2002) found that providers for TNR colonies mostly thought that the 'colonies they fed had an excellent or good quality of life'. This study also pointed out, though, that there was a 33% death or disappearance rate in these colonies over a median period of 18 months. Equally, Kalz *et al.* (2000) noted a 33% survival rate over a 42-month period – that is, only one-third of cats survived 3.5 years. But then, statistics on death and disappearance from TNR colonies tell us little about living cats' welfare.

More positively, a complex study of TNR colonies on a Florida university campus found that only 4% of cats were euthanized for humane reasons. Of the cats remaining at the end of the 7-year study, 83% had been there for more than 6 years. These cats were generally in 'adequate physical condition', with no significant weight difference from owned cats (Levy *et al.*, 2003). But these colonies had high rates of adopting out, and the cats had a constant provision of hygienically supplied food. So, it would be unwise to generalize from the Levy *et al.* (2003) study to all other TNR colonies, where conditions might be less good.

However, at least some cats in TNR colonies, especially where the colony is supported and cats are vaccinated, appear to have good or adequate welfare and a reasonable lifespan, albeit shorter than the average lifespan of owned cats. There may, of course, be some TNR colonies where cats do have poor welfare; but this does not seem to be generally the case. However, as I will suggest later, there is a difficulty here: what would count as an acceptable level of welfare for feral cats? Should the comparison in terms of welfare or longevity be with homed cats or, instead, with wild animals living in similar areas, such as racoons or red foxes?

9.4.3 Feral cats destroy wildlife

The third argument for restricting or eliminating feral cat populations is that they have a negative impact on wildlife. Here, I will consider cats only as predators, though recent research also emphasizes cats' role as disease vectors for wildlife (Hess, 2011; Jessup and Miller, 2011). It is worth noting, too, that owned cats with outdoor access also contribute to wildlife predation; some influential studies are derived from owner reports about their own cat's predation (most prominently, Woods *et al.*, 2003). Where TNR colonies and homed cats with outdoor access are located close to one another, differentiating their wildlife impact is very difficult.

Cats are skilled hunters, favouring small mammals but also catching birds and reptiles. There is significant evidence that where introduced on oceanic islands, they have a major impact on wildlife (Nogales *et al.*, 2004). This is especially true where prey species – such as rabbits – have also been introduced, creating a phenomenon called hyperpredation. This occurs when populations of the introduced prey species allow the population of the predator species to remain higher than it otherwise would, were only native species available. For instance, rabbits may provide cats with food over the winter when migrating birds are absent and the cats would otherwise go hungry. The resulting higher cat population increases the rate of predation on native as well as introduced prey species, diminishing their numbers very quickly (Courchamp *et al.*, 1999).

While there is general agreement about the effects of cat predation on oceanic islands, cats' impacts on continental wildlife and in urban/suburban areas are more hotly contested. Some conservationists argue that, given widespread human development, urban parks and gardens have become critical habitats for bird species, especially for migrants. Woods *et al.* (2003) argue that homed cats with outdoor access could cause hyperpredation, where food provision by cats' owners allows cats to exist in much greater density than if they were dependent on predation. This raised density increases cat predation on wildlife, as cats hunt whether hungry or not, and experimental evidence indicates that they prioritize hunting over eating even preferred foods (Adamec, 1976). As many TNR colonies are also subsidized by humans, the same argument about hyperpredation looks likely to hold for TNR (Longcore *et al.*, 2009).

Defenders of feral cats and of TNR policies resist this characterization of the effects of feline predation in most cases (and corresponding proposals for restraints on homed cats having outdoor access; Burrows, 2004; Lepczyk *et al.*, 2010). They argue that human development, vehicular trauma and migratory bird collisions with human structures are a much greater threat. A focus on cats, they argue, distracts from the indiscriminate killing and displacing of wildlife by humans. And there are cases where cat predation keeps populations of other species that compete with or prey on native wildlife under control, such as rats, though admittedly these cases are relatively few (see, for instance, Tidemann *et al.*, 1994).

It is difficult to know how to assess the evidence of cat predation on wildlife, as only a small number of studies exist, with very few focused on TNR colonies. There is evidence that where feral cats are located near to endangered native species of ground-nesting birds or mice, cat predation has a significant effect (Hawkins *et al.*, 1999). But evidence of a threat is less strong where wildlife habitat is already significantly disturbed, for instance in suburban gardens or on university campuses. In these areas, Dickman (2009, p. 45) may be right that the 'direct impact of cats on native fauna will be secondary to the more dramatic effects of loss and modification of native vegetation by the suburbs themselves' (though Dickman was specifically discussing Australian house cats; geographical location may matter here). This is not to say that cats do not kill a lot of wildlife; there is strong evidence that they do (Woods *et al.*, 2003). But in urban/suburban areas, most of those killed are not members of threatened species or populations.

So, concerns about public health, cat welfare and wildlife conservation underpin arguments that feral cat populations should be reduced. Although a few people reject all such concerns, most agree that some control of feral cat populations is needed. This leads to consideration of the effectiveness of TNR and TE in doing so.

9.5 Is TNR Effective? How Does it Compare with TE?

To judge the effectiveness of TNR, we need to know what 'effectiveness' means here. Many of those who support TNR maintain that its aim is to create a stable or declining cat population (e.g. Slater, 2007). Suppose this is the aim. Is TNR effective in achieving it?

Clearly, if cats that would have reproduced freely are neutered, there will be fewer cats than otherwise. An extensive 2006 study of cats sterilized across a number of different US states concluded: 'With a pregnancy rate of 0.9 [litters/year/female cat], a litter size of 4.1, and a female proportion of 53.4%, the sterilization of 103,643 cats by the TNR programs could be expected to prevent the birth of 204,224 kittens in a breeding season' (Wallace and Levy, 2006, p. 281). Even though there is normally a high 6-month feral kitten mortality rate, TNR programmes mean that many fewer cats exist than would otherwise have been the case. However, that fewer cats come into existence does not equate to population stabilization or reduction; it may just mean that populations increase at a slower rate. Does TNR actually stabilize or reduce population size?

Stoskopf and Nutter (2004) compared six sterilized, maintained feral cat colonies in North Carolina with three maintained but unsterilized control colonies. While the number of cats in the sterilized colonies underwent a mean decrease of 36% over 2 years, the number in the control colonies increased 47%. Over time, one of the six sterilized colonies became extinct and one neared extinction, while four appeared to stabilize at lower numbers. Stoskopf and Nutter estimated that it would take 4–10 years for a sterilized feral cat colony to die out. In another study (Levy *et al.*, 2003), cats in TNR colonies on the University of South Florida campus decreased over 10 years from 155 to 23. However, this was a TNR+ colony, with a high level of 'adopting out', so may not be typical.

Cat colonies can increase in size in two ways: by reproduction and by immigration. TNR only prevents reproduction. Some claim that established cat colonies defend their territory against incomers, meaning that a TNR colony, once established, should only reduce in size (Scott *et al.*, 2002a). But there is substantial contradictory evidence. Studies consistently show cats moving into and between colonies. Levy *et al.* (2003) explicitly note this, and Zaunbrecher and Smith (1993) report an established TNR colony of 30 cats being joined by six new cats over time. (This may contrast with stray dogs: see Molento, Chapter 7, this volume.) In addition, the visible existence of TNR colonies may encourage people to dump their unwanted cats there. One study, for instance, found that well-fed cat colonies encouraged the illegal abandonment of cats (Castillo, 2001). Abandoned cats are also often not neutered, which means that an ongoing programme of neutering is needed in TNR colonies; it is estimated that a rate of between 71% and 94% of cats must be neutered for feral cat populations to decline (Andersen *et al.*, 2004; Foley *et al.*, 2005). But, as Jessup (2004) notes, most TNR programmes are volunteer based, understaffed and cannot constantly sterilize all cats in a colony. So, high abandonment rates at TNR colonies could increase feral cat numbers.

In summary, TNR programmes mean that fewer cats exist than in uncontrolled cat colonies; colonies may stabilize or decline in population. Whether and how fast they decline depends on immigration, ongoing sterilization and the adopting out of socialized cats. But TNR can be effective, if 'effective' means 'achieving a stable or declining population'.

But effectiveness depends on the goal. Longcore *et al.* (2009) maintain that feral cat advocates have different goals from wildlife conservationists. Conservationists aim to eliminate free-roaming cat predation on wildlife, by eliminating either the roaming or the cats. In the case of unadoptable feral cats, where roaming cannot be eliminated, TE – eliminating the cats – not TNR is the obvious alternative. With TE, unlike TNR, cats are removed permanently; they not only are unable to reproduce but also are unable to predate. From this perspective, only TE (or another form of elimination) is effective.

It is sometimes argued that TE itself does not, in practice, reduce feral cat numbers. Scott *et al.* (2002a) maintain that: 'Methods that remove cats permanently often result in a new group of cats moving in to fill the void, unless the cats are in a geographically restricted area, such as an island.' However, it is difficult to find evidence to support this claim. Levy *et al.* (2003) reported that completely vacated cat colonies were not recolonized by other cats, even where food remained available. And an empty colony is unlikely to encourage the cat abandonment that a thriving colony encourages. But, none the less, if existing feral cats did move in, and if these cats reproduced, then the wildlife advantage of TE would be less clear-cut.

For those who accept the need to control feral cat populations – as most people do – a key question, then, is whether to stabilize and (probably) reduce cat populations slowly by TNR; to reduce such populations swiftly and substantially by TE; or to use different methods in different cases. How these alternatives are judged depends a good deal on the values to which individuals are committed.

9.6 Values at Stake

Underlying these various cat management tools are different human values. These include the lives of living beings (including humans, cats and wildlife) and the subjective states of these living beings, such as enjoyment or suffering, excitement or fear. On some views, the performance of 'natural' behaviours and possession of certain kinds of freedoms are valuable, independently of any accompanying subjective experiences. The continued existence of particular species of wildlife, or of certain forms of biodiversity, is sometimes argued to be critically important, either in itself or in terms of instrumental value to people or ecosystems. And a variety of different human/non-human relationships may be of moral relevance here. TNR proposals affirm or prioritize certain of these values, while TE proposals prioritize others; this lies at the heart of the moral debate. Here, I will explore just three potential values 'at stake': the value of life, the value of subjective experience/animal welfare and the value of species. Concerns about 'naturalness' and human relations also recur throughout. Although public health concerns are also important here, they are less prominent in debates between defenders of TNR and TE, perhaps because managed TNR colonies can resolve some public health issues (also see Molento, Chapter 7, this volume, for a related discussion of public health).

9.6.1 The value of lives

TNR policies imply (i) that the lives of any cats that actually exist are valuable and should be protected, but that (ii) either allowing more lives to be created is not in itself valuable or that such value is outweighed by likely negative effects: (i) provides a reason for preferring TNR over TE, while (ii) justifies preferring

TNR over doing nothing (and may suggest some common ground with advocates of TE).

Why should we protect cats' lives? This question is not about suffering: euthanizing cats, if done well, should be painless. Rather, the concern is whether depriving a cat of a longer life than it could otherwise have had is wrong if that life would have been reasonably good. This raises substantial questions about whether cats have a sense of themselves as persisting over time; whether painless killing could be against their interests; and how their interests should feature in our moral decision making. See Franco *et al.* (Chapter 4, this volume) and Palmer (2010) for a more extended discussion.

Certainly, from several leading theoretical positions in animal ethics, killing animals even painlessly is prima facie morally problematic. On Regan's (2004) animal rights view, all adult mammals have inherent value and basic rights, including the right not to be killed. Others argue that animals, including cats, have (in some sense) a desire to go on living and that this desire should not be frustrated, even if the animals concerned do not actually experience this frustration (Singer, 1993; Francione, 2006). The idea that painlessly killing sentient animals harms them is widely accepted in animal ethics (though on some views this does not mean it is impermissible, if there are benefits that outweigh the harm). Although the claim that painless killing causes harm can reasonably be contested, I will not question this, as moral concern about killing cats lies at the heart of the conflict about TNR versus TE.

However, it is not only cats' lives that are at stake. It is difficult to argue – on the basis of cats' capacities alone, anyway – that their lives have special value lacked by their prey. Any ethical position that values the lives of mammals would also have to count the lives of rodents as well as cats; and many additionally value bird lives (Regan, 2004). ACA (2012b), for instance, claims that: 'We value the relationship between people, the earth, and all animals and acknowledge that the inherent interests of all sentient beings must be given equal consideration.'[3] But it is not obvious how endorsing TNR does give 'equal consideration' to all the sentient lives at stake, as statistics indicate that each cat takes many animal lives (Woods *et al.*, 2003).

One approach here is to deny that lives lost can be summed – to say that there is nothing in principle morally worse about the loss of (say) 20 rabbit lives than one cat life (Taurek, 1977). But this view is difficult to defend, and even if one did take this view, if the cat life were always favoured, the claims of 'equal consideration' would seem hollow.

There are, however, several possible ethical arguments that would protect cats' lives while tolerating cats' predation on other lives, arguments not based solely on consequentialist reasoning (where what matters is outcomes in terms of numbers of deaths):

1. Adult humans are moral agents, morally responsible for their actions; they can be praised and blamed for what they do. Cats are not moral

agents. If a human kills a cat, that person is morally responsible for the death of the cat (perhaps violating its right to life). But as cats are not moral agents, what they do falls outside the moral realm. So, it is not a matter of moral concern whether cats kill rabbits (or mice, or birds). Thus, TNR is permissible but TE is not.

2. Feral cats have become naturalized in ecosystems; they are wildlife. We do not intervene in other kinds of predation, such as when lions prey on antelope; why pick out cats? What happens in ecosystems either is not our moral business (so there is no requirement to intervene) or (more strongly) we should not intervene in natural processes such as predation. This view is compatible with treating all lives with equal consideration.

3. Humans (or some humans) have special moral responsibilities to at least some unowned cats that they do not have to wildlife, either because cats are, in some sense, community members or because of a historical, 'backward-looking' human responsibility generated by breeding cats and then abandoning them. These special relationships mean we should assist unowned cats, not harm them. So, while TNR is permissible or even desirable, TE is unacceptable.

Although there is some plausibility in each of these arguments, an advocate of TE over TNR has a number of possible responses.

A defender of TE might say of (1): first, this view implies that we do not have moral responsibilities to assist others, in this case wildlife. But on many moral views, we should assist others who are threatened, even if the threats do not come from a moral agent. If I should rescue a child from a rabid dog, should I not also protect a bird from a cat? Second, the view expressed in (2) is sometimes called 'agent centred'. It focuses on what the moral agent does, not on what the consequences will be. If we prioritize the outcome, rather than the agent, TNR results in (say) 20 dead rabbits and one live cat, while TE results in one dead cat and 20 live rabbits. Surely, the killing of 20 sentient animals is worse than the killing of one. So, if I do not kill the cat, I am responsible for allowing 20 rabbits to be killed that I otherwise could have saved; my inaction has permitted a much worse outcome. (This response has a different basis in ethical theory; the focus is on the consequences, not the agent.)

The 'naturalness' argument – (2) above – is resisted strongly by wildlife conservationists. They maintain that cats in TNR colonies are often subsidized, so they are to this degree unnatural, and that this means their populations are unnaturally dense compared with other predators. In addition, in cases where cats are not native, prey species may not be adapted to their hunting techniques. So, any argument that we should not intervene in natural systems to remove feral cats fails, because their ecosystem impacts are much greater than would be the case with natural predators. If we think that naturalness – understood as being (in some sense) not of human origin (e.g. Elliot, 1982) – is valuable, then there is something right about this argument. Certainly, it is hard to see how all-neutered, subsidized TNR colonies could be regarded as a 'natural part of the landscape', a claim ACA (2012a) makes about feral cats. But then, most (but not all) TNR colonies are themselves located in environments that have been altered substantially by people, including in urban and suburban areas, where little is natural anyway.

The final arguments here (3) concerned the possibility of humans' special obligations to unowned cats, either because domesticated cats are members of an extended mixed human/domestic animal community (Midgley, 1983) or because we have special responsibility to unowned cats based on human past mistreatment and abandonment of them (Palmer, 2010). Both arguments imply that cats have a special moral 'pull' on us. However, it is not clear why this moral pull should not be extended to urban wildlife as well. Birds that are attracted to gardens intentionally by nest boxes and feeders could also be viewed as part of a mixed human/animal community; and if people put cats in situations where they threaten wildlife, then presumably special obligations to the wildlife are also generated, because humans are now responsible for threats to wildlife as well as for the abandonment of cats. Neither of these approaches, in many cases anyway, will give reasons to protect cats' lives while tolerating their predation.

Now I will turn to the second value at stake – subjective animal welfare.

9.6.2 Welfare value

Welfare has many interpretations; I will take it to refer to animals' positive or negative subjective experiences, where 'positive welfare' includes feelings of pleasure, satisfaction or excitement and 'negative welfare' feelings of pain, suffering, distress or frustration. As Keeling *et al.* (2011) note, welfare in this sense concerns the state of the individual animal, not our moral obligations towards it. Poor welfare does not automatically generate a duty to assist; what, if anything, should be done depends on what one values and how such values fit into one's ethical theory. For the purposes of this section, however, I shall take good welfare to be valuable. See Franco *et al.* (Chapter 4, this volume) for further discussion.

A key issue in the debate here is what counts as morally acceptable welfare for feral cats. A common question is whether feral cats have a life worth living, here taking this to mean a life in which the intrinsically good states outweigh the intrinsically bad states (McMahan, 2009). This could form some kind of baseline: if a cat would be better off dead, and death was the only realistic alternative to a miserable life, this would be an argument for TE. As noted, though, many cats in TNR colonies, especially where subsidized and vaccinated, certainly have a life worth living in this sense. A further question, though, concerns comparative welfare. The short lifespan of cats in TNR projects, for instance, is often compared unfavourably to homed cats and taken as evidence of poor welfare. A short lifespan, though, does not give us clear information about welfare preceding death. And – if we change the comparison class – the survival rates of feral cats, even those outside TNR projects, are comparable to those of wildlife. For instance, the 25% survival rate of feral kittens is similar to the survival rate of juvenile red foxes in urban/agricultural areas (a recent Illinois study placed this at 24%; Gosselink *et al.*, 2007). Although, like cats, captive red foxes can live for 14 years, their average lifespan in the wild is only 1.5 years. By this measure, feral cats in TNR colonies survive fairly well. This raises the question whether those who argue that feral cats have unacceptably short lives and poor welfare also think this about wild animals, and that we should consider euthanizing them; or, alternatively, whether they think that different standards apply to feral cats.

Let us now turn to the cats' prey animals. Here, we are interested less in longer-term welfare than in the effects of cat–wildlife interactions, often at the end of a wild animal's life. It is very likely that cats inflict intense pain, and in some cases drawn-out suffering, on the animals and birds they successfully hunt. Ethical concerns about this are raised by recent papers objecting to TNR policies, such as Longcore *et al.* (2009) and Lepczyk *et al.* (2010). Jessup (2004, pp. 1377, 1378) claims that: 'Free-roaming and feral cats yearly kill hundreds of millions, perhaps as many as a billion, native North American birds, mammals, reptiles, amphibians, and fish', noting that they are 'maimed, mauled, dismembered, ripped apart, and gutted while still alive, and if they survive the encounter, they often die of sepsis because of the virulent nature of the oral flora of cats'. Cats also may generate disturbance, change bird behaviour and cause negative states such as alarm and fear in the animals they hunt but fail to catch.

How should concerns about cat and wildlife welfare and suffering figure in thinking about TNR versus TE? TNR protects the lives of cats, many of whom have lives worth living. It also allows cats to experience the positive welfare brought about by hunting experiences. TE will deprive those cats of their lives, and positive welfare, but protects wildlife from reduced welfare caused by cats' hunting activity.

Utilitarians would aggregate values such as pleasure and pain here. At first sight, on this approach, TE gives us a world with less negative experience from predation (and more positive experience from continuing wild lives) but (assuming the cats also would have had lives worth living) without the positive values of the cats' future lives. In contrast, TNR would allow a good deal of negative wild subjective experience and the loss of the positive experiences the wildlife would otherwise have had; but the cats continue to live and gain positive hunting experiences. But making this kind of calculation is difficult. Different cat colonies in different situations will generate different sums. And what will happen to wildlife not caught by cats? Suppose that – as seems plausible – on at least some occasions, cats catch weak or sick wild animals; cat predation might reduce overall suffering, or may mean that surviving wild animals face less competition for scarce resources, so have better welfare.

There is a further factor here, too. Worries about wild suffering surely extend further than cat predation. McMahan (2010) raises the following concern: 'Suppose that we could arrange the gradual extinction of carnivorous species, replacing them with new herbivorous ones. Or suppose that we could intervene genetically, so that currently carnivorous species would gradually evolve into herbivorous ones ... If we could bring about the end of predation by one or the other of these means at little cost to ourselves, ought we to do it?' Of course, there are currently practical hurdles to doing this. But if we object to feral cats on the grounds of the suffering they cause, then theoretically we may be committed to much broader ideas of intervention in the wild – which may itself be incompatible with conserving wildlife.

Arguments that we should maximize positive values net of negative values, such as suffering, have been important – if largely implicit – in the TNR/TE debate. Given the diverse situations of feral cats, and their access to wildlife, taking this approach seems to require a case-by-case analysis; in some cases, TE might be expected to maximize positive values, and in other cases, TNR will be more likely to do so. No general conclusion seems possible on the basis of aggregated concerns about positive and negative welfare.

9.6.3 Species value

Wildlife conservationists argue in favour of TE not just because cats kill and cause suffering to wildlife, but because – they claim – cat predation is affecting species populations negatively, and in some cases driving species to extinction. This is an intense concern where feral cat colonies are close to wildlife refuges; but, as noted earlier, it is also argued that migratory bird populations are seriously impacted. These arguments usually just assume that species are of significant value. But, given that the need to protect species populations is offered as a compelling reason for killing sentient cats, such arguments would – from perspectives that value cats' lives and welfare, at least – need to be very convincing.

Different kinds of arguments about the value of species are possible. It is sometimes maintained that a species is better understood as a kind of individual than as a group or class, and that it makes sense to say a species has interests. This claim then serves as the ground for an attribution of moral status to species (Johnson, 2003). However, arguments that species have moral status in a similar way to individuals are difficult to justify and have been widely rejected (Sarkar, 2005; Sandler and Crane, 2006). Others claim that species, or biodiversity, have some kind of 'intrinsic value'; but again, the arguments here are difficult to substantiate (for a thoroughgoing critique of such arguments, see Maier, 2012).

This leaves arguments about the value of species preservation to humans and to ecosystem function (and therefore indirectly to humans). Some species are important resources; some are aesthetically significant (including some migrating bird species threatened by cats); some are 'keystone' or 'load bearing' in ecosystems. The loss of some species may make other species more vulnerable, leading to cascading extinctions (Norton, 1987). These reasons underpin the view that the protection of species and biodiversity is critical for present and future human welfare, and perhaps for the welfare of future sentient animals. So, even though cats may primarily endanger species that are not obviously of use to us now, such species may play a role in significant ecological cycles that we do not currently understand. So, TE advocates maintain, we should prioritize these values over the lives of feral cats now and adopt TE over TNR.

Defenders of TNR, though, have a variety of responses available. Most do not deny that cats have eliminated some species and contribute to threats to others. But they argue that human activity is a much greater threat to species than cats. The Intergovernmental Panel on Climate Change (IPCC, 2007, p. 19) estimates with 'medium confidence' that climate change (independent of invasive species and habitat loss) puts '20–30% of species assessed so far at increased risk of extinction if increases in global average warming exceed 1.5–2.5°C over 1980–1999 levels'. This is far greater than the effect of cats; cats merely add minor pressures to the major ones imposed by people; but (a TNR defender may respond) they risk becoming scapegoats, at least to a minor degree.

Even if cats do threaten species, though, TNR defenders can argue that this does not justify killing cats (though it can justify preventing them from being born). From a rights perspective, species themselves are not sentient or intelligent individuals and have no moral status. Species may be instrumentally valuable; but the instrumental value of a species could not outweigh the rights of individual cats not to be killed by moral agents. A weaker related response could accept that cats may increase the risk of species extinction but maintain that the choice here is between an enhanced risk of species extinction and certain killing of cats. Given this tough choice, we should avoid the certain negative of killing cats and accept the risk of extinctions.

The conflict between prioritizing species values/biodiversity or the value of individual animals' lives forms yet another part of the moral conflict about TNR or TE. Protecting species may require killing cats, and protecting cats risks losing culturally, and possibly ecologically, important species. Different values therefore point in different policy directions. A further complication arises here because the TNR/TE debate is just one element of a broader, growing discussion about what conservation means in the context of climate change and pervasive human environmental impacts. Should we any longer attempt to preserve so-called 'natural' ecosystems and endangered species? How far should we come to terms with living in, and with, novel ecosystems (Marris, 2011)? Should we just leave cats alone and see how things shake out, instead of trying to fix the 'problems' we have created and thereby potentially generating new ones? The fast-moving nature of these debates makes it difficult to draw firm conclusions about the role of feral cats in ecosystems, or TNR versus TE.

9.6.4 Drawing threads together

In this chapter, I have tried to unpack the key values, especially concerning animal lives, animal welfare and wildlife conservation, that underpin debates about TE/TNR. Few people consider these values a matter of complete indifference, but, as I hope to have suggested, how important these values are thought to be and how they are prioritized varies widely. Although this chapter has been exploratory, it has, I think, highlighted factors it seems important to consider when policy making about cat management. Those making decisions about particular unowned cat colonies might, for instance, consider questions such as the following:

Public health
Does the existence of this cat colony present a public health risk? If so, how serious is the risk – are human lives threatened? Could the risk be reduced or removed by TNR and associated practices such as vaccination and deworming? Are there ways in which public health can be protected without taking the lives of cats?

Cat welfare
Do the cats in this location have a life worth living? Are there affordable and feasible ways of improving their health around the TNR process? Are there cats in this location that have a reasonable prospect of a better life if adopted out? Are these cats in a location where they are at high risk or likely to undergo significant suffering (due to climate, a major road, a hostile local human population)? Is there a plausible future in which these cats will have enough shelter and food to live reasonable lives – can a reliable carer or, better, group of carers be identified? Is this colony in a place that is likely to attract people to dump non-neutered cats into it? If so, can it be located anywhere else, or are there other ways of dissuading people from dumping? Is there a long-term prospect that these cats, if neutered, will have a reasonable quality of life and that the colony will stabilize or decline in size?

Wildlife conservation
What do we know about wildlife in the area in which this colony is located? How urban/suburban/wild is this colony's location, and how disturbed is it already by human activity? What are the likely food sources for these cats, if not provided by human carers? How much native wildlife is located within the

cats' likely hunting zone, and how much of it is ground nesting? Is this colony located near to a wildlife reserve? Which migrating bird species pass through this area, how endangered are they and how vulnerable do studies show members of these bird species are to cats? Are there ways of reducing these cats' access to wildlife? Can the cat colony be moved if the cats appear to be exerting high wildlife pressure?

Of course, these are by no means the only relevant questions here; and being able to answer them requires a consideration of each feral cat colony on its own terms, rather than the adoption of a 'one-size-fits-all' policy. Answering questions of this kind will not necessarily make what to do clear, anyway; while a cat colony may not threaten public health, and cat welfare may be good, local wildlife may be seriously threatened, meaning that values point in different directions and require prioritization. But still, taking questions of this kind seriously at least means that the key values relevant to each situation are not overlooked.

9.7 Conclusions

- Most agree that feral cat populations should be managed, for cat welfare, public health and ecological reasons, though the severity of these problems is debated.
- Methods of management are contested, in particular (the focus here) whether to adopt policies of trap–neuter–return (TNR) or trap–euthanize (TE).
- TNR programmes can stabilize/cause decline in feral cat populations. But conservationists argue that TNR is too slow to protect wildlife, preferring TE.
- TNR and TE raise many questions about value: of lives, subjective welfare, naturalness, species/biodiversity and certain kinds of human/animal relationships.
- Questions are also raised in ethical theory: should we refrain from harm/protect rights or try to bring about the best consequences, however interpreted?
- Advocates of TNR generally:
  - emphasize the value of cats' lives
  - claim that feral cats in TNR colonies have reasonable welfare and killing them (even painlessly) is wrong
  - distinguish morally between humans killing cats and cats killing wildlife
  - maintain that humans are a greater threat to the environment than cats.
- Advocates of TE generally:
  - argue that euthanizing feral cats reduces overall suffering (both of feral cats themselves and wildlife)

  - maintain that killing feral cats saves more lives overall (a consequentialist argument), as well as protecting ecological and species values
  - claim that feral cats are invasive or unnatural.
- Some disagreement here reflects conflicting interpretations of empirical questions.
- Some values are shared but differently prioritized. This makes for a genuine moral conflict: there are serious moral reasons to adopt competing policies.
- This debate is located in a broader, unresolved nexus of issues about the meaning of conservation in the context of global environmental change.
- Policies about feral cat management should be contextual and negotiated locally, and at least consider value questions relating to public health, cat welfare and wildlife conservation.

Questions for discussion

1. You care for a local trap–neuter–return (TNR) colony and these cats often come into your garden. Should you remove your bird feeder? Should you encourage your neighbours to remove theirs?
2. You are carrying out TNR on a colony of non-neutered cats. One of them tests positive for FelV but appears to be healthy and to have a life worth living. What should you do?
3. A small group of feral cats has recently become established near to a local shoreline where you know there to be endangered ground-nesting piping plovers. You hear of a proposal to create a TNR colony and your neighbours volunteer to provision it. What should you do?
4. A critic of the creation of TNR colonies argues that feral cats should be treated as wildlife and therefore we should neither feed them nor neuter them. How do you respond to this argument?

Acknowledgements

The author gratefully acknowledges useful comments provided on earlier versions of this chapter by T.J. Kasperbauer and by Mike Appleby, Peter Sandøe, Dan Weary and the other contributors to this volume. In addition, thanks are due to participants who contributed to a fruitful discussion of an earlier version of this chapter at the Yale University Interdisciplinary Center for Bioethics.

Notes

[1]To keep a tight focus here, subsidiary ethical problems such as whether TNR schemes should identify and euthanize FIV+ (Feline AIDS positive) cats will not be

discussed, nor related questions such as whether cat owners should allow their cats to go outdoors.

[2]The term, 'trap–euthanize' will be adopted, as it is standard terminology, while recognizing that, strictly speaking, 'euthanasia' should be for the good of the one that dies, and that is not necessarily the case here.

[3]And (though to develop this would open up another issue) homed cats and cats in TNR colonies are fed on a diet of other animals often killed by people for this purpose.

References

ACA (Alley Cat Allies) (2012a) Cats and the environment resource center (http://www.alleycat.org/page.aspx?pid=324, accessed 17 September 2012).

ACA (2012b) About us/values (http://www.alleycat.org/page.aspx?pid=616, accessed 16 September 2012).

Adamec, R.E. (1976) The interaction of hunger and preying in the domestic cat (*Felis catus*): an adaptive hierarchy. *Behavioral Biology* 18, 263–272.

Afonso, E., Thuliez, P. and Gilot-Fromont, E. (2006) Transmission of *Toxoplasma gondii* in an urban population of domestic cats (*Felix catus*). *International Journal for Parasitology* 36, 1373–1382.

Andersen, M.C., Martin, B.J. and Roemer, G.W. (2004) Use of matrix population models to estimate the efficacy of euthanasia versus trap–neuter return for management of free-roaming cats. *Journal of the American Veterinary Medical Association* 225, 1871–1876.

ASPCA (American Society for the Prevention of Cruelty to Animals) (2012) Position statement on feral cat management (http://www.aspca.org/About-Us/policy-positions/feral-cat-management, accessed 27 August 2012).

Burrows, P. (2004) Professional, ethical and legal dilemmas of trap-neuter-release. *Journal of the American Veterinary Medical Association* 225, 1365–1369.

Castillo, D. (2001) Population estimates and behavioral analyses of managed cat colonies located in Miami-Dade County, Florida Parks. MSc thesis, Florida International University, Miami, FL.

Centoze, L. and Levy, J. (2002) Characteristics of free-roaming cats and their caretakers. *Journal of the American Veterinary Association* 220, 1627–1633.

Courchamp, F., Langlais, M. and Sugihara, G. (1999) Control of rabbits to protect island birds from cat predation. *Biological Conservation* 89, 219–225.

Cuffe, D.J. (1983) Ear-tipping for identification of neutered feral cats. *Veterinary Record* 112(6), 129.

Dickman, C.R. (2009) Housecats as predators in the Australian environment: impacts and management. *Human–Wildlife Conflicts* 3(1), 41–48.

Elliot, R. (1982) Faking Nature. *Inquiry* 25, 81–93.

Foley, P., Foley, J.E., Levy, J.K. and Paik, T. (2005) Analysis of the impact of trap–neuter return programs on populations of feral cats. *Journal of the American Veterinary Medical Association* 227(11), 1175–1181.

Francione, G. (2006) Equal consideration and the interest of animals in continuing existence: a response to Professor Sunstein. *University of Chicago Legal Forum* 231, 239–240.

Gosselink, T., Van Deelen, T., Warner, R. and Mankin, P. (2007) Survival and cause-specific mortality of red foxes in agricultural and urban areas of Illinois. *Wildlife Management* 71(6), 1862–1873.

Hart, B.L. and Barrett, R.E. (1973) Effects of castration on fighting, roaming, and urine spraying in adult male cats. *Journal of the American Veterinary Medical Association* 163, 290–292.

Hawkins, C.C., Grant, W.E. and Longnecker, M.T. (1999) Effects of subsidized house cats on California birds and rodents. *Transactions of the Western Section of the Wildlife Society* 35, 29–33.

Hess, S.C. (2011) By land and by sea. *Wildlife Professional* 5(1) 66–67.

HSUS (Humane Society of the United States) (2011) US pet ownership statistics (http://www.humanesociety.org/issues/pet_overpopulation/facts/pet_ownership_statistics.html, accessed 31 July 2012).

IPCC (Intergovernmental Panel on Climate Change) (2007) *Climate Change: The AR4 Synthesis Report*. IPCC, Geneva, Switzerland.

Jessup, D.A. (2004) The welfare of feral cats and wildlife. *Journal of the American Veterinary Medical Association* 225, 1377–1383.

Jessup, D. and Miller, M. (2011) The trickle down effect: how toxoplasmosis from cats can kill sea otters. *Wildlife Professional* 5(1), 62–65.

Johnson, L. (2003) Future generations and contemporary ethics. *Environmental Values* 12(4), 471–487.

Kalz, B., Shiebe, K.M., Wegner, I. and Priemer, J. (2000) Health status and causes of mortality in feral cats in a delimited area of the inner city of Berlin. *Berliner und Münchener tierärztliche Wochenschrift* 113, 417–422.

Kass, P. (2005) Cat overpopulation in the United States. *Animal Welfare* 3, 119–139.

Keeling, L., Rushen, J. and Duncan, I.J.H. (2011) Understanding animal welfare. In: Appleby, M.C., Mench, J.A., Olsson, I.A.S. and Hughes, B.O. (eds) *Animal Welfare*, 2nd edn. CAB International, Wallingford, UK, pp. 13–26.

Lepczyk, C., Dauphine, D., Bird, D., Conant, S., Cooper, R., Duffy, D., *et al*. (2010) What conservation biologists can do to counter trap–neuter–return: response to Longcore *et al*. *Conservation Biology* 24(2), 627–629.

Levy, J.K., Gale, D.W. and Gale, L.A. (2003) Evaluation of a long-term trap neuter return and adoption program on a free-roaming cat population. *Journal of the American Veterinary Association* 222, 42–46.

Longcore, T., Rich, C. and Sullivan, L. (2009) Critical assessment of claims regarding management of feral cats by trap–neuter–return. *Conservation Biology* 23(4), 887–894.

McMahan, J. (2009) Assymmetries in the morality of causing people to exist. In: Roberts, M. and Wasserman, D. (eds) *Harming Future People: Ethics, Genetics and the Non-Identity Problem*. Springer, Dordrecht, the Netherlands, pp. 49–70.

McMahan, J. (2010) The meat-eaters. *New York Times*, 19 September (http://opinionator.blogs.nytimes.com/2010/09/19/the-meat-eaters/, accessed 17 December 2012).

Maier, D. (2012) *What's So Good About Biodiversity?* Springer, Dordrecht, the Netherlands.

Marris, E. (2011) *Rambunctious Garden*. Bloomsbury, London.

Midgley, M. (1983) *Animals and Why they Matter*. Georgia University Press, Athens, GA.

Mott, M. (2004) U.S. faces growing feral cat problem. *National Geographic News* (http://news.nationalgeographic.com/news/2004/09/0907_040907_feralcats.html, accessed 17 September 2012).

Nogales, M., Martin, A., Tershy, B.R., Donlan, C.J., Veitch, R., Puerta, N., *et al.* (2004) A review of feral cat eradication on islands. *Conservation Biology* 18, 310–319.

Norton, B. (1987) *Why Preserve Natural Variety?* Princeton University Press, Princeton, NJ.

Palmer, C. (2010) *Animal Ethics in Context*. Columbia University Press, New York.

PETA (People for the Ethical Treatment of Animals) (2012) Animal rights uncompromised: feral cats (http://www.peta.org/about/why-peta/feral-cats.aspx, accessed 17 September 2012).

PFMA (Pet Food Manufacturers Association) (2011) Pet population (http://www.pfma.org.uk/pet-population-2011/, accessed 16 September 2012).

Regan, T. (2004) *The Case for Animal Rights*, 2nd edn. University of California Press, Berkeley, CA.

Sandler, R. and Crane, J. (2006) On the moral considerability of *Homo sapiens* and other species. *Environmental Values* 15(1), 69–84.

Sarkar, S. (2005) *Biodiversity and Environmental Philosophy*. Cambridge University Press, Cambridge, UK.

Scott, K., Levy, J. and Crawford, P. (2002a) Characteristics of free-roaming cats evaluated in a trap–neuter–return program. *Journal of the American Veterinary Association* 221, 1136–1138.

Scott, K.C., Levy, J.K. and Gorman, S.P. (2002b) Body condition of feral cats and the effect of neutering. *Journal of Applied Animal Welfare Science* 5, 203–213.

Singer, P. (1993) *Practical Ethics*, 2nd edn. Cambridge University Press, Cambridge, UK.

Slater, M. (2007) The welfare of feral cats. In: Rochlitz, I. (ed.) *Animal Welfare*. Springer, Dordrecht, the Netherlands, pp. 141–175.

Stoskopf, M.K. and Nutter, F.B. (2004) Analyzing approaches to feral cat management – one size does not fit all. *Journal of the American Veterinary Medical Association* 225, 1361–1364.

Taurek, J. (1977) Should the numbers count? *Philosophy and Public Affairs* 6(4), 293–316.

Tidemann, C.R., Yorkston, H.D. and Russack, A.J. (1994) The diet of cats, *Felis catus*, on Christmas Island, Indian Ocean. *Wildlife Research* 21, 279–286.

Wallace, J.A. and Levy, J.K. (2006) Population characteristics of feral cats admitted to seven trap–neuter–return programs in the United States. *Journal of Feline Medicine and Surgery* 8, 279–284.

Woods, M., McDonald, R. and Harris, S. (2003) Predation of wildlife by domestic cats *Felis catus* in Great Britain. *Mammal Review* 33(2), 174–188.

Zaunbrecher, K.I. and Smith, R.E. (1993) Neutering feral cats as an alternative to eradication programs. *Journal of the American Veterinary Medical Association* 203(3), 449–452.

10

Alone or Together: A Risk Assessment Approach to Group Housing

Jeff Rushen* and Anne Marie de Passillé
University of British Columbia, Canada

10.1 Abstract

The question of whether it is better to house animals in groups rather than alone deals with unresolved issues at the heart of animal welfare. In particular, we need to be able to rank different classes of threats to welfare (e.g. behavioural deprivation versus illness) on a single scale. In this chapter, we discuss the risk assessment approach to animal welfare that is currently being developed and ask to what extent it helps in answering this question. Qualitative risk assessment has been used to compare group housing and individual housing for dairy cows, and the conclusions are that while the hazards of tie stalls are more serious than those for loose housing on deep straw, the high incidence of lameness in cows housed in free stalls can result in a greater overall risk to welfare than occurs in tie stalls. Thus, the details of the particular group housing used need to be considered. We also discuss whether a quantitative risk assessment is possible and conclude that, as yet, too little is known about the magnitude of different threats to animal welfare, and there remains too much disagreement over the fundamental nature of animal welfare for this to be currently feasible.

*E-mail: rushenj@mail.ubc.ca

10.2 Introduction

> There are known unknowns; that is to say we know there are some things we do not know. But there are also unknown unknowns – there are things we do not know we don't know.
>
> (Donald Rumsfeld, US Secretary of Defense, 2002)

The question of whether or not it is better for an animal to be housed in a social group rather than alone deals with unresolved issues at the heart of animal welfare. These include the issue of how best to balance different threats to animal welfare, how precisely we can measure overall welfare as opposed to the components of welfare and how best to define welfare so as to enable an overall welfare assessment. Within the current context of animal welfare, the question of individual versus group housing is particularly relevant for lactating cows, pregnant pigs, dairy and veal calves, horses, fur animals and laboratory animals. Generally, individual housing is promoted because it reduces the occurrence of aggression and social competition, and the transmission of illness between animals. Group housing is promoted as a means of allowing social contact and other natural behaviours, and usually results in animals having more space.

It is not our intention to carry out a thorough review to propose the best answer, but instead we will use this issue to explore some of the weaknesses with our current concepts of animal welfare and to explore what we need to know in order to answer the question more definitively. In the spirit of Donald Rumsfeld, we will try to move some 'unknown unknowns' into the realm of the 'known unknowns'. First, we tackle the issue of how best to judge the relative impact of different threats to animal welfare. We discuss the qualitative risk assessment approach to animal welfare of the European Food Safety Authority (EFSA), which promises a more principled method of placing threats to welfare in order of priority. Although the issues we raise are pertinent for all species, we shall explore the underlying issues by focusing on the question of whether dairy cows are better off in loose housing (especially free-stall/cubicle housing – Fig. 10.1 – and straw yards) or tie stalls (Fig. 10.2). We adopt this focus because EFSA recently reviewed research on the welfare of dairy cows and conducted a risk assessment, which may help answer this question (EFSA 2009a,b,c,d,e,f). Second, we examine whether we have a clear enough concept of animal welfare and sufficient data to conduct a more quantitative risk assessment, using some of the 'unknowns' about the relationship between aggression and animal welfare to demonstrate that it is appropriate to be wary of simple assumptions.

10.3 A Risk Assessment Approach to Animal Welfare

This section focuses on the issue of how best to judge the relative impact of different threats to animal welfare.

Fig. 10.1. Free-stall housing (or cubicle housing) is the most commonly used indoor group housing for dairy cows.

Fig. 10.2. An example of tie-stall housing for dairy cows.

EFSA (www.efsa.europa.eu) acts as an independent source of scientific advice to support European policies and legislation associated with the food chain. EFSA's remit covers the traditional areas of food safety, nutrition and animal health, and has a mandate to deliver scientific opinions on animal welfare (Serratosa and Ribó, 2009). In this area, EFSA aims to provide objective and independent science-based advice using a formal risk assessment procedure, which was modelled on that used to assess risks to food safety and animal health established by the World Organization for Animal Health (OIE) and in the *Codex Alimentarius* (Smulders, 2009). EFSA has published a number of opinions on the welfare of animals. An important part of EFSA's mandate is to develop an 'objective' methodology for placing various threats to welfare in some order of priority, so that limited resources (governmental or private) can be directed towards the animal welfare issues where they are most needed (Algers, 2009).

The EFSA risk assessment process is described in a number of publications (e.g. Algers, 2009; Smulders, 2009; EFSA, 2012; Müller-Graf *et al.*, 2012). The process is an evolving one that has its own specialized terminology, which also changes as the process evolves; this makes a succinct description challenging. However, the underlying concepts remain relatively constant. The purpose of risk assessment is to rank the various threats to welfare (or hazards) in some order of priority. A hazard is defined as a 'factor/condition with a potential to cause a negative animal welfare effect (or adverse effect)' (Müller-Graf *et al.*, 2012, p. 233). For the purposes of animal welfare assessment, these hazards are usually aspects of the housing or management of the animals, such as wet bedding, infrequent hoof trimming or high density. More recently, EFSA has recognized that many housing and management factors can have a beneficial effect on animal welfare and now tends to speak of 'factors' which can have either negative effects (hazards) or positive effects (EFSA, 2012). For the sake of simplicity, in this chapter we will consider only 'hazards'. The risk assessment process is multi-step:

> Risk assessment is a scientifically based process that seeks to determine the likelihood and consequences of an adverse event, which is referred to as a hazard. It generally consists of the following steps: (i) hazard identification, (ii) hazard characterization, (iii) exposure assessment and (iv) risk characterization.
>
> Müller-Graf *et al.* (2012, p. 232)

For the purposes of this chapter, we will focus on the stage of hazard characterization, which has been labelled more recently as 'consequence characterization' (EFSA, 2012).

Hazard characterization aims to measure the nature and seriousness of the adverse effects on animal welfare that result when an animal is exposed to a particular hazard. To do this, scores are given for the 'severity' (sometimes called 'intensity') of the adverse effect, based on the consequences of the animal's exposure to the hazard. These are then multiplied by the duration of the adverse effects, to give a score which represents the magnitude of the consequences of exposure to the hazard for an individual animal that does suffer the adverse consequences. Not all animals will experience the same adverse effects when exposed to a hazard. For example, large, dominant animals are less likely to experience negative effects from overcrowding than are small, timid ones. Therefore, the magnitude score is then multiplied by an estimate of the proportion of animals that will experience the adverse effects if exposed to the hazard (which is referred to as a likelihood score). The final product measures the magnitude of the consequences of exposure to the hazard when applied to a group of animals. Ideally, the risk assessment would be based on quantitative measures of severity, duration and likelihood, derived from research. However, sufficient data are rarely available to do this, and so a qualitative risk assessment is used in which the measures are derived from expert opinion.

An important point of this analysis is that the impact of any hazard on animal welfare is not only a function of its severity but also of its duration, which is often overlooked in discussions of animal welfare. A hazard with low severity may have a larger overall impact on animal welfare than one with a high severity if it lasts much longer. Similarly, the fact that not all animals will suffer from the adverse consequences when exposed to a hazard is also underappreciated. For example, while illness may appear to be a more severe hazard than behavioural deprivation, the proportion of the animals that become ill may be much smaller than those that suffer from behavioural problems. These issues are discussed in more concrete detail later.

With this brief description of the outlines of the risk assessment approach, we will turn to our concrete example: are the consequences for the welfare of dairy cows of exposure to the hazards that are associated with group housing (free stalls and straw yards, Fig. 10.1) greater or lesser than those associated with individual housing (tie stalls, Fig. 10.2)?

10.4 The EFSA's Qualitative Risk Assessment: Group or Individual Housing for Lactating Cows

EFSA has published an extensive report on the welfare of dairy cows (EFSA, 2009f), as well as scientific opinions on the most serious hazards of dairy cows, using the qualitative risk assessment procedure (EFSA, 2009a,b,c,d,e). The risk assessment was done separately for tie stalls, free stalls, straw yards and pasture, and separately for leg and locomotor problems, udder problems, metabolic and reproductive problems and for problems involving behaviour, pain and fear. The opinions also grouped the hazards into those dealing with housing, feeding and nutrition, management and genetic selection. Table 10.1 is a summary of the number of hazards identified and the magnitude of the

Table 10.1. The number of hazards and the summed magnitude of the adverse effects of the hazards in each type of housing system for dairy cows that were identified by the EFSA following a qualitative risk assessment.

	Tie stalls	Free stalls	Straw yards
Leg and locomotor problems			
Number of hazards	22	29	24
Summed magnitude of hazards			
Housing	493	710	335
Feeding	188	188	143
Management	315	338	330
Genetics	75	75	75
Summed overall	1071	1311	883
Udder problems			
Number of hazards	28	30	28
Summed magnitude of hazards			
Housing	51	59	44
Feeding	3	3	3
Management	55	62	130
Genetics	10	10	10
Summed overall	119	134	187
Metabolic and reproductive problems			
Number of hazards	48	54	51
Summed magnitude of hazards			
Housing	239	542	245
Feeding	242	242	242
Management	120	193	122
Genetics	155	155	155
Summed overall	756	1132	764
Behaviour, fear and pain problems			
Number of hazards	43	49	44
Summed magnitude of hazards			
Housing	1629	1321	662
Feeding	271	272	271
Management	150	224	208
Genetics	125	125	130
Summed overall	2175	1942	1271
Total			
Number of hazards	141	162	147
Summed magnitude of hazards			
Housing	2412	2632	1286
Feeding	704	705	659
Management	640	817	790
Genetics	365	365	370
Summed overall	4121	4519	3105

adverse effects associated with each grouping of hazards for tie stalls, free stalls and straw yards. As the issues associated with keeping cows at pasture are very different from those associated with indoor housing, this chapter is focused on the latter.

A simple way to determine the relative advantage of individual housing (tie stalls) and group housing (free stalls and straw yards) is to examine the summed magnitude scores for all of the hazards. Table 10.1 shows that the 141 hazards present in tie stalls have a summed magnitude score slightly lower than the 162 hazards in free stalls, but substantially higher than the score of the 147 hazards in straw yards. Therefore, tie stalls represent a greater threat to animal welfare than straw yards, but free stalls are a slightly greater threat than tie stalls. The first conclusion, then, is that the answer to the question of this chapter (whether it is better to keep animals in individual or group housing) depends on the type of group housing. Some group housing systems are better for the cows' welfare than individual housing, while others are not.

The most serious hazards facing cows in tie stalls were judged to be those that resulted in behavioural problems, fear or pain (Table 10.1). One of the most serious hazards was 'inadequate bedding', with a magnitude score of 270. The severity of this hazard was judged as 3 on a scale of 4. However, the high overall magnitude score came from the long duration of the effects (365 days a year for zero-grazed animals) and the large percentage of the animals that was affected (>70%) (EFSA, 2009a). In contrast, straw yards did worse than tie stalls in hazards leading to udder problems, such as mastitis (Table 10.1). 'Inadequate bedding', leading to systemic mastitis and teat injury, was an important hazard, also with a severity of 3. However, the overall magnitude score in this case was much lower than in tie stalls (7.5), because of the short duration of the adverse effect of mastitis (<20 days) and the smaller number of animals involved (<40%) (EFSA, 2009c). This emphasizes the importance of estimates of duration and likelihood in expert judgement of the impact of any hazard on animal welfare.

Examining Table 10.1 shows that the major disadvantages of free stalls compared to tie stalls are the hazards associated with housing leading to leg and locomotor problems and metabolic and reproductive problems. The main advantages of straw-pack housing come from the fact that the magnitudes of these hazards are lower than in free stalls. This conclusion is supported by data which tend to show that the occurrence of foot lesions and lameness is higher in free stalls than in tie stalls (Sogstad *et al.*, 2005; Cramer *et al.*, 2008) and in straw yards (Haskell *et al.*, 2006). There are many fewer data available on the occurrence of metabolic disorders. However, the incidence of ketosis is lower in free stalls than in tie stalls (Schnier *et al.*, 2004; Simensen *et al.*, 2010). This illustrates a fairly obvious but often underappreciated point: that the disadvantages of any group housing system are not solely those associated with aggression, and that judging the relative welfare advantage of two housing systems must take into account overall animal welfare rather than a single factor. One strength of the risk assessment approach is that it attempts to list all hazards associated with different housing systems.

One important question concerns the extent that the hazards associated with a particular housing system can be controlled through changes in management or whether they are an intrinsic part of the housing system and cannot be changed easily. In general, hazards associated with genetics and housing are more difficult for farmers to control than are hazards associated with management and feeding. For example, the largest hazard magnitudes for free stalls resulted from having too few stalls for the number of cows (EFSA, 2009b,d). However, this hazard can be managed relatively easily by reducing the number of cows. For tie stalls, the biggest hazards involved the behavioural consequences of poor stall design, of inadequate bedding and of being tied without exercise. While the second may be managed by improving the management of bedding, reducing the first would require expensive changes to buildings (or buying smaller cows), while the third cannot really be managed at all. An imperfect but simple way of getting a rough answer to the degree to which the problems in any housing system can be managed is to calculate the ratio of the hazards identified by EFSA as being linked to management compared to those linked to housing. This ratio is 0.27 for tie stalls, 0.31 for free stalls and 0.61 for straw yards. Thus, it can be concluded that the problems with group housing (especially for straw yards) are controlled more easily through better management (without major capital costs) than are the problems associated with tie stalls.

This qualitative risk assessment therefore answers the question posed in the title of this chapter: the hazards associated with individual housing in tie stalls are higher than those associated with group housing in straw yards, even though aggression may be a problem in the latter. Furthermore, the hazards of group housing are controlled more easily by management. Use of free stalls, however, can reduce the advantages of group housing, mainly from an increase in foot and leg health problems, but these could be improved through better management.

The limitation with a qualitative assessment, however, is that the scores for severity, duration and likelihood are based on expert opinion, and the question can always be raised as to whether or not the experts are correct. The following sections examine what still needs to be known to carry out a more quantitative risk assessment.

10.5 Quantitative Risk Assessment

To quantify the severity or intensity of any hazard, some clear definition of animal welfare is needed. If the aim of the risk assessment is to place various hazards on a list of priority, then the relative severity of different hazards must be on a single linear scale, which can be applied to all types of hazards. This assumes that there is a single 'currency' that can be used to rate the severity of different hazards, which raises the questions of what exactly is animal welfare and how risk assessors can agree on a common definition of welfare that can be applied in all cases (Algers, 2009).

EFSA is neither clear nor consistent on the scale that should be used to rate severity. For example, Müller-Graf *et al.* (2012) base their assessments of severity on the degree to which an animal experiences 'pain, malaise, fear or anxiety', while other writings refer more to 'pain and distress' (EFSA, 2012, p. 13) or 'pain and suffering' (p. 23). In other cases, however, the intensity is defined as the increasing intensity of a number of different measures or else an increase in the number of measures, e.g. optimum social behaviour, physical comfort, likelihood of death, etc. (EFSA, 2012, p. 19). The important point for this chapter, however, is that regardless of how severity is defined, to rank different hazards in order of importance, the severity of the consequences of the different hazards must, at some point, be compared using a single scale. To achieve this, a clear definition of welfare is needed.

In the 1980s and 1990s, there was lively debate among scientists as to the best definition of animal welfare (Nordenfeldt, 2006; Fraser, 2008). Most of the disagreements occurred because the various proponents were defending definitions of welfare that aimed to describe the essence of animal welfare in terms of a single characteristic. The most famous, perhaps, is the view that animal welfare is all about animal feelings, especially the extent that the animal 'suffers' (Duncan and Petherick, 1991; Dawkins, 2008; see Weary, Chapter 11, this volume). In this view, the relative impact of individual or group housing on animal welfare will depend on how much animals suffer as a result of being housed in these ways. The advantage with this type of definition is that it does, in theory at least, provide us with a common currency with which we can weight or rank the different threats to animal welfare. The question of whether individual housing is better or worse for animal welfare than being the victim of aggression is answered by determining which causes the most suffering.

Unfortunately, over the decades there has been disagreement between scientists about the scientific credibility of claims about animal emotions, sentience or consciousness. There is not space in this chapter to deal in detail with the debate about the scientific standing of animal consciousness. However, while some scientists continue to express scepticism, many, if not most, scientists working in animal welfare now accept that animal sentience is a valid topic for scientific investigation, at least in principle (e.g. Dawkins, 2008). However, even if we accept in principle that animal emotions or sentience can be studied scientifically, there remain grave practical problems when we attempt to quantify the amount of suffering that real animals endure in any given housing system. Most successful attempts to measure suffering in animals have focused on relatively short-lasting emotions, such as those associated with pain resulting from acute procedures such as dehorning, branding and tail docking (Weary *et al.*, 2006). We argue that there have been no successful attempts to measure the degree of long-term suffering associated with a method of housing, although the methods of determining cognitive bias hold some promise for this (Mendl *et al.*, 2009).

However, the emotional states of animals must be considered in defining animal welfare, not only because of their importance in the biology of animals but also because this is what people are most concerned about. Fraser *et al.* (1997) reminded us that scientists did not become interested in the concept of animal welfare because of any particular scientific findings or because it best explained some biological phenomenon. Rather, it was because of very vocal concerns about the way we treated animals, especially farm and laboratory animals, being expressed by many members of the public. Clearly, in order to address public concerns, we must know what these concerns are, and our definition of animal welfare must deal with them, otherwise the research we do becomes irrelevant to the underlying societal issue. The desire to address public concerns has led to the increased use of definitions that are based on the notion that animal welfare is inherently multidimensional (e.g. Fraser, 2008) and simply list all of the proposed elements without necessarily placing them in some order of priority. The most well-known example is that of the Five Freedoms, which define good welfare as approaching the ideal state of 'freedom' from hunger and thirst, from discomfort, from pain, injury and disease and from fear and distress, and 'freedom' to express normal behaviour.

The advantage of such definitions is their pragmatism, in that it is easier to obtain consensus from the diverse stakeholders, as nobody's concerns are left out. A disadvantage of such definitions is that they do not allow us to weight or rank the different threats to welfare. For example, the definition based on the Five Freedoms does not allow us to say whether the increased 'freedom' to express normal social behaviour, which would occur when animals are placed in groups, would outweigh the reduction in the 'freedom' from pain, injury or disease that might result from increased aggression or illness in groups.

Thus, these definitions of animal welfare make it difficult to answer the question of this chapter, which is whether individual housing is worse for animal welfare than the hazards present when animals are housed in groups. To quantify the severity of any particular hazard with sufficient precision to allow a quantitative risk assessment, we need a definition of animal welfare that allows us to rank hazards on a single scale.

10.6 Quantitative Risk Assessment: What Types of Data Do We Need?

In this section, we ask whether we have sufficient data available (and the right sort of data) to carry out a quantitative risk assessment on group versus individual housing for dairy cattle. We do this by first considering health issues, and then turn to behavioural issues.

10.6.1 Comparing the welfare consequences of various diseases

We come closest to having adequate data to assess the severity, duration and likelihood of an adverse effect on animal welfare for threats to welfare that arise from diseases. These data can be used to make some judgement as to the relative impact on welfare of the diseases that occur with different prevalence in tie stalls and free stalls.

Because of their economic importance, there are fairly good data concerning the duration and incidence of various diseases affecting dairy cattle. Furthermore, and over the past few decades, there has been interest in finding common behavioural changes that occur following illness, and these common changes may allow us to quantify the severity. The concept of 'sickness behaviour' holds that all forms of illness can result in some common behavioural changes (Weary *et al.*, 2009). The most common behavioural response to illness is reduced feeding behaviour, and many diseases of cattle result in reduced feed intake (Bareille *et al.*, 2003). Therefore, feed suppression following an illness may be used as one indicator of the severity of the illness (Rushen *et al.*, 2008).

Bareille *et al.* (2003) examined diseases affecting lactating cows on a free-stall farm in Switzerland and measured the declines in feed intake that resulted from these diseases. We recognize that it may not be possible to extrapolate the values to other farms or other countries, but the results serve as an example of how the risk assessment approach could be used to assess the relative impact of different diseases on animal welfare. Table 10.2 shows data from Bareille *et al.* (2003) for ten of the most common diseases. The table shows the estimated severity, which is based on the average daily reduction in feed intake, and the relative ranking of the diseases. This is, of course, a rough estimate since the severity of the disease will vary over time, but it is useful for the purposes of illustration. The table shows the duration and the product of duration and severity, which allows a second ranking. What is noticeable is that the relative importance of the diseases changes. For example, milk fever, which was ranked second in importance on the basis of severity alone, dropped to fifth place because of the relatively short duration of the effect. In contrast, ketosis, which was ranked fourth in terms of severity alone, moved to top place because of the long duration of the effect.

The last three columns show how the rankings change further when the incidence of the diseases is considered. A quantitative risk assessment requires an estimation of the percentage of animals that are exposed to a risk factor that then suffer from the relative adverse effects. If it is assumed that free-stall housing is a risk factor for these various diseases, the measures of the incidence of each disease can be taken to assess this. The table shows that further changes in rank occur when the number of animals suffering from the disease is added. For example, a very difficult calving dropped from fourth place to eighth because few animals suffered from this. In contrast, systemic mastitis increased because it was one of the most common maladies affecting the cows.

Table 10.2. The severity (based on reduction of daily feed intake), duration and incidence of different diseases affecting dairy cattle. The table shows the products of the variables and the relative rankings of the diseases. Data taken from Bareille *et al.* (2003).

Disease	Severity (reduction in feed intake kg/day)	Rank	Duration (days)	S × D	Rank	Incidence (per cent of animals)	S × D × I	Rank
Ketosis	0.55	4	131	71.9	1	24.2	1739	1
Lameness (hocks)	1.56	1	31	48.1	2	18.3	880	2
Metritis	0.32	7	145	46.8	3	12.8	599	3
Milk fever	0.85	2	45	38.2	5	14.7	562	4
Systemic mastitis	0.26	9	117	30.2	7	16.0	483	5
Lameness (feet)	0.47	5	60	27.8	8	8.9	247	6
Very difficult calving	0.62	3	70	43.4	4	5.5	239	7
Difficult calving	0.38	6	98	37	6	6.3	233	8
Retained placenta	0.19	10	56	10.4	9	13.9	145	9
Local mastitis	0.27	8	6	1.6	10	16.0	26	10

Note: S = severity; D = duration; I = incidence.

Thus, data on the severity, duration and incidence of health disorders can help identify the most serious threats to welfare. Can this approach be extended to compare free stalls and tie stalls? Again, some data are available to help us do this, particularly from a series of studies in Norway, where many farms have moved from tie-stall housing to loose housing. This work has shown that, for example, the incidence of ketosis is lower in free stalls than in tie stalls (1.30 versus 3.39 occurrences/cow year) (Schnier *et al.*, 2004; Simensen *et al.*, 2010). In contrast, the prevalence of hoof lesions was lower in tie stalls than in free stalls (48% versus 72%; Sogstad *et al.*, 2005), as has been reported in other countries (Cramer *et al.*, 2008). By taking the figures of Bareille *et al.* (2003) for the product of severity × duration for ketosis and lameness due to hoof lesions (and assuming that the prevalence of lameness is the same as the incidence), the total hazard magnitude (for these two diseases) for tie stalls can be calculated to be 1578 and for free stalls to be 2095. The final result reflects the high prevalence of hoof lesions in free stalls compared to that of ketosis in tie stalls. Obviously, this comparison is only valid for comparing the housing systems in terms of the occurrence of ketosis and lameness. But, it illustrates the type of data needed to compare the impact of different diseases on animal welfare in different housing systems.

10.6.2 Unknowns when comparing the welfare consequences of behavioural problems: judging the severity of aggression

Can a quantitative risk analysis approach to behavioural-based hazards associated with group or individual housing be developed? To do so, accurate estimates would be needed of the severity, duration and likelihood for all of the behavioural consequences of being housed in groups or individually. Unfortunately, there is still a lack of sufficient knowledge and data to do this, which is illustrated by briefly reviewing the research that has been carried out on one behavioural consequence of group housing: aggression. Is enough understood about the relationship between aggression and animal welfare to quantify the severity of welfare effects from increased aggression? In this section, we consider some of the evidence that the link between aggression and animal welfare may be more complex than we often assume. To do so, we need to consider other species as well, as there has been little research on aggressive behaviour in dairy cows.

Aggressive behaviour is often considered to be wholly negative and all efforts are made to reduce the occurrence of aggression and social competition. However, in taking this view, the biological complexity of aggressive behaviour and its implications for animal welfare have been underestimated. In particular, research has suggested that, in a number of ways, aggressive behaviour can be considered as a sign of good welfare.

First, aggression is clearly a 'natural' behaviour, with clear functional benefits to the aggressor in at least some circumstances, such as increased access to resources (Mitani *et al.*, 2010; Lindenfors and Tullberg, 2011). Second, many animals seem highly motivated to behave aggressively and success in aggressive encounters is rewarding (e.g. Fuxjager *et al.*, 2010) and, for some humans at least, pleasurable; young boys report that fighting is an enjoyable activity (Benenson *et al.*, 2008). There is some evidence that highly aggressive animals may experience a rebound in aggressive motivation if prevented from fighting (Kudryavtseva *et al.*, 2011), suggesting that they may suffer from deprivation of aggressive behaviour.

Aggressive cattle (Fig. 10.3) may also enjoy collateral benefits: cattle selected for fighting ability are generally less fearful and appear more resistant to some stressful situations (Plusquellec and Bouissou, 2001). Aggression may also be a sign of a healthy cow: during the week before calving, cows diagnosed with severe metritis (after calving) engaged in fewer aggressive interactions over feed than cows that remained healthy (Huzzey *et al.*, 2007). Thus, according to most definitions of animal welfare, being aggressive and winning an aggressive encounter would improve welfare, and the occurrence of success in aggression could be considered as a positive welfare indicator.

Fig. 10.3. Aggressive behaviour between large animals such as cattle often appears dramatic and we usually assume that such behaviour has a negative impact on animal welfare. However, we lack good data on the actual magnitude of the effect of aggression on animal welfare. Furthermore, in some cases, it can be argued that the occurrence of some forms of aggression can have a positive effect on animal welfare, especially the more playful types of aggression illustrated here.

Obviously, though, aggressive encounters rarely end with two winners and the losers can suffer from a variety of negative consequences, and winners can also suffer injury. It is because of these negative consequences that the aim is to reduce aggression between animals, but this assumes that the negative consequences outweigh the positive consequences. However, there is rarely empirical evidence that this is the case.

What are the actual consequences of defeat? For some species, these are sufficiently severe that death follows. But are such extreme consequences true for all animals? In a study of silver fox vixens' willingness to work (i.e. pull a chain) to gain access to a shared cage which held a conspecific, Hovland *et al.* (2008) noted that initial encounters would sometimes result in overt aggression, but this did not appear to stop the vixens continuing to visit the shared cage, nor were the results strongly affected by which vixen was dominant or subordinate. Interestingly, the pairs differed greatly in how much aggression they showed, and a later study (Hovland *et al.*, 2011) found some evidence that the time the vixens spent together was greater if their social interactions were positive. However, what was notable was that the aggression did not stop the vixens from working for social contact. This suggests that, for these animals, the experience of aggression and of subordinate status is not highly aversive. Val-Laillet *et al.* (2009) also noted that most displacements between cattle occurred between individuals that spent most time in proximity.

These effects may reflect some post-conflict resolution, in which two animals are able to 'make up' after an incident of aggression and maintain their relationship. This has long been known to occur among primates, and is an important mechanism underlying group cohesion. Unfortunately, other than a recent study in horses (Cozzi *et al.*, 2010), there has been little research on this topic in farm animals.

Thus, the relationship between aggression and animal welfare may be more complex than has been assumed, and measures of the severity of exposure to aggression may need to be refined.

10.7 What Do We Need to Carry Out a Quantitative Risk Assessment?

Thus, to carry out a quantitative risk analysis of the relative impacts of group versus individual housing requires that we have a definition of welfare that is sufficiently precise to assess the severity of different classes of hazard. Some means are available to assess the relative severity and overall impact of different diseases on the welfare of dairy cows, but this involves only a subset of diseases. We need more systematic collection of data to assess the severity, duration and likelihood of animals suffering from behavioural problems, and a clearer concept of animal welfare that allows us to incorporate some of the complexity of behavioural threats to animal welfare, such as those arising from aggression.

10.8 Conclusions

- Risk analysis holds promise as a way of assessing the overall impact of hazards associated with housing on animal welfare.
- The impact of any particular hazard on animal welfare will depend not only on its severity but also on its duration and the proportion of animals within a group that suffer from the adverse consequences.
- A qualitative risk assessment has concluded that the hazards of individual housing (tie stalls) pose more of a threat to the welfare of dairy cattle than those of some group housing systems (open, bedded packs).
- The welfare advantages of group housing, however, can be lost when free-stall housing is associated with increased incidence of leg and locomotor problems.
- The relative advantages and disadvantages of group housing require that we look at the impacts of all hazards associated with actual housing systems, rather than focusing on aggression alone.
- The development of quantitative risk analysis requires that we have a clearer definition of animal welfare that allows us to assess the relative severity of different classes of hazards.
- To quantify the effects of behavioural problems on animal welfare, we need a better method of assessing severity and more systematic collection of data on the duration of the effect and the proportion of individuals that are affected.

Questions for discussion

1. Can we rank different classes of threat to animal welfare (e.g. disease, behavioural deprivation) objectively or does the relative importance of each inevitably depend on the personal values of individual people?
2. If you were reborn as a dairy cow, would you prefer to be kept in a free-stall barn or a tie-stall barn? Explain why.
3. Under what conditions could we say that experiencing aggression is good for animal welfare?

References

Algers, B. (2009) A risk assessment approach to animal welfare. In: Smulders, F.J.M. and Algers, B. (eds) *Welfare of Production Animals: Assessment and Management of Risks*. Wageningen Academic Publishers, Wageningen, the Netherlands, pp. 223–237.

Bareille, N., Beaudeau, F., Billon, S., Robert, A. and Faverdin, P. (2003) Effects of health disorders on feed intake and milk production in dairy cows. *Livestock Production Science* 83, 53–62.

Benenson, J.F., Carder, H.P. and Geib-Cole, S.J. (2008) The development of boys' preferential pleasure in physical aggression. *Aggressive Behavior* 34, 154–166.

Cozzi, A., Sighieri, C., Gazzano, A., Nicol, C.J. and Baragli, P. (2010) Post-conflict friendly reunion in a permanent group of horses (*Equus caballus*). *Behavioural Processes* 85, 185–190.

Cramer, G., Lissemore, K.D., Guard, C.L., Leslie, K.E. and Kelton, D.F. (2008) Herd- and cow-level prevalence of foot lesions in Ontario dairy cattle. *Journal of Dairy Science* 91, 3888–3895.

Dawkins, M.S. (2008) The science of animal suffering. *Ethology* 114, 937–945.

Duncan, I.J. and Petherick, J.C. (1991) The implications of cognitive processes for animal welfare. *Journal of Animal Science* 69, 5017–5022.

EFSA (European Food Safety Authority) (2009a) Scientific opinion of the Panel on Animal Health and Welfare on a request from the Commission on the risk assessment of the impact of housing, nutrition and feeding, management and genetic selection on behaviour, fear and pain in dairy cows. *EFSA Journal* 1139, 68pp.

EFSA (2009b) Scientific opinion of the Panel on Animal Health and Welfare on a request from the Commission on the risk assessment of the impact of housing, nutrition and feeding, management and genetic selection on metabolic and reproductive problems in dairy cows. *EFSA Journal* 1140, 75pp.

EFSA (2009c) Scientific opinion of the Panel on Animal Health and Welfare on a request from the Commission on the risk assessment of the impact of housing, nutrition and feeding, management and genetic selection on udder problems in dairy cows. *EFSA Journal* 1141, 60pp.

EFSA (2009d) Scientific opinion of the Panel on Animal Health and Welfare on a request from the Commission on the risk assessment of the impact of housing, nutrition and feeding, management and genetic selection on leg and locomotion problems in dairy cows. *EFSA Journal* 1142, 57pp.

EFSA (2009e) Scientific opinion of the Panel on Animal Health and Welfare on a request from the European Commission on the welfare of dairy cows. *EFSA Journal* 1143, 38pp.

EFSA (2009f) Scientific report of EFSA prepared by the Animal Health and Animal Welfare Unit on the effects of farming systems on dairy cow welfare and disease. *Annex to the EFSA Journal* 1143, 284pp.

EFSA (2012) Guidance on risk assessment for animal welfare. *The EFSA Journal* 10, 2513, 30pp.

Fraser, D. (2008) *Understanding Animal Welfare.*Wiley-Blackwell, Chichester, UK.

Fraser, D., Weary, D.M., Pajor, E.A. and Milligan, B.N. (1997) A scientific conception of animal welfare that reflects ethical concerns. *Animal Welfare* 6, 187–205.

Fuxjager, M.J., Forbes-Lorman, R.M., Coss, D.J., Auger, C.J., Auger, A.P. and Marler, C.A. (2010) Winning territorial disputes selectively enhances androgen sensitivity in neural pathways related to motivation and social aggression. *Proceedings of the National Academy of Sciences* 107, 12393–12398.

Haskell, M.J., Rennie, L.J., Bowell, V.A., Bell, M.J. and Lawrence, A.B. (2006) Housing system, milk production, and zero-grazing effects on lameness and leg injury in dairy cows. *Journal of Dairy Science* 89, 4259–4266.

Hovland, A.L., Mason, G.J., Kirkden, R.D. and Bakken, M. (2008) The nature and strength of social motivations in young farmed silver fox vixens (*Vulpes vulpes*). *Applied Animal Behaviour Science* 111, 357–372.

Hovland, A.L., Akre, A.K., Flø, A., Bakken, M., Koistinen, T. and Mason, G.J. (2011) Two's company? Solitary vixens' motivations for seeking social contact. *Applied Animal Behaviour Science* 135, 110–120.

Huzzey, J.M., Veira, D.M., Weary, D.M. and Von Keyserlingk, M.A.G. (2007) Prepartum behavior and dry matter intake identify dairy cows at risk for metritis. *Journal of Dairy Science* 90, 3220–3233.

Kudryavtseva, N., Smagin, D.A. and Bondar, N.P. (2011) Modeling fighting deprivation effect in mouse repeated aggression paradigm. *Progress in Neuro-Psychopharmacology and Biological Psychiatry* 35, 1472–1478.

Lindenfors, P. and Tullberg, B.S. (2011) Evolutionary aspects of aggression: the importance of sexual selection. In: Huber, R., Bannasch, D.L. and Brennan, P. (eds) *Advances in Genetics. Volume 75: Aggression*. Academic Press, Boston, pp. 7–22.

Mendl, M., Burman, O.H.P., Parker, R.M.A. and Paul, E.S. (2009) Cognitive bias as an indicator of animal emotion and welfare: emerging evidence and underlying mechanisms. *Applied Animal Behaviour Science* 118, 161–181.

Mitani, J.C., Watts, D.P. and Amsler, S.J. (2010) Lethal intergroup aggression leads to territorial expansion in wild chimpanzees. *Current Biology* 20, R507–R508.

Müller-Graf, C., Berthe, F., Grudnik, T., Peeler, E. and Afonso, A. (2012) Risk assessment in fish welfare, applications and limitations. *Fish Physiology and Biochemistry* 38, 231–241.

Nordenfelt, L. (2006) *Animal and Human Health and Welfare*. CAB International, Wallingford, UK.

Plusquellec, P. and Bouissou, M.-F. (2001) Behavioural characteristics of two dairy breeds of cows selected (Hérens) or not (Brune des Alpes) for fighting and dominance ability. *Applied Animal Behaviour Science* 72, 1–21.

Rushen, J., de Passillé, A.M.B., von Keyserlingk, M.A.G. and Weary, D.M. (2008) *The Welfare of Cattle*. Springer, Dordrecht, The Netherlands.

Schnier, C., Hielm, S. and Saloniemi, H.S. (2004) Comparison of the disease incidences of Finnish Ayrshire and Finnish Black and White dairy cows. *Preventive Veterinary Medicine* 62, 285–298.

Serratosa, J. and Ribó, O. (2009) International context and impact of EFSA activities in animal welfare in the European Union. In: Smulders, F.J.M. and Algers, B. (eds) *Welfare of Production Animals: Assessment and Management of Risks*. Wageningen Academic Publishers, Wageningen, the Netherlands, pp. 275–304.

Simensen, E., Østerås, O., Bøe, K.E., Kielland, C., Ruud, L.E. and Næss, G. (2010) Housing system and herd size interactions in Norwegian dairy herds; associations with performance and disease incidence. *Acta Veterinaria Scandinavica* 52, 14 (http://www.actavetscand.com/content/52/1/14, accessed 20 August 2013).

Smulders, F.J.M. (2009) A practicable approach to assessing risks for animal welfare – methodological considerations. In: Smulders, F.J.M. and Algers, B. (eds) *Welfare of Production Animals: Assessment and Management of Risks*. Wageningen Academic Publishers, Wageningen, the Netherlands, pp. 239–274.

Sogstad, Å.M., Fjeldaas, T., Østeras, O. and Forshell, K.P. (2005) Prevalence of claw lesions in Norwegian dairy cattle housed in tie stalls and free stalls. *Preventive Veterinary Medicine* 70, 191–209.

Val-Laillet, D., Guesdon, V., von Keyserlingk, M.A.G., de Passillé, A.M. and Rushen, J. (2009) Allogrooming in cattle: relationships between social preferences, feeding displacements and social dominance. *Applied Animal Behaviour Science* 116, 141–149.

Weary, D.M., Niel, L., Flower, F.C. and Fraser, D. (2006) Identifying and preventing pain in animals. *Applied Animal Behaviour Science* 100, 64–76.

Weary, D.M., Huzzey, J.M. and Von Keyserlingk, M.A.G. (2009) Board-invited review: Using behavior to predict and identify ill health in animals. *Journal of Animal Science* 87, 770–777.

What is Suffering in Animals?

11

Daniel M. Weary*
University of British Columbia, Canada

11.1 Abstract

In much of the recent animal welfare literature the word 'suffering' is used simply as an adjunct (as in 'pain and suffering') or to emphasize that the animal consciously perceives pain or some other negative affect. A stronger usage of the term implies that the negative feelings are prolonged, high intensity or both, but without any clear line to distinguish when suffering begins. Researchers in human medicine have developed more explicit definitions of suffering that also reference concurrent negative feelings (including fear, anxiety, sadness and depression) and the patient's ability to cope. Applying this broader definition of suffering to animal welfare will require a new approach to the research we do. Research on animal suffering will require not only the assessment of negative affective states but also an assessment of how concurrent negative states interact, a general assessment of the animal's emotional health and its ability to cope with adversity.

11.2 Introduction

The word 'suffering' comes with moral loading, suggesting an extra responsibility for action and explaining why it is often used in the rhetoric of animal advocates and in criminal law relating to animal care (e.g. 'causing unnecessary

*E-mail: dan.weary@ubc.ca

suffering' as specified in section 445.1 of the Criminal Code of Canada). Thus, a clear understanding of suffering is central to animal welfare, but few scholars have addressed the issue of animal suffering explicitly, and those that have often seem to use the word simply as an adjunct (as in 'pain and suffering'), providing little guidance for how actually to assess or prevent suffering. In contrast, the human medical literature has delved deeply into how patients conceive of their own suffering and offers a range of methods for assessing and preventing suffering in practice. This chapter provides a brief summary of this literature and suggests how we might adapt these ideas from the medical literature for application to animals.

11.3 Current Usage

In the scientific literature on animal welfare, the word 'suffering' is typically used in conjunction with the experience of some negative affective state. Most often, this affective state is pain, as is clear from many of the examples used below, but suffering is also used in conjunction with other negative feelings, including fear.

The simplest usage of the term is adjunctive, as in 'pain and suffering', without attempt to distinguish the two ideas (e.g. Elwood, 2011; EFSA, 2012). I suggest that this use is essentially meaningless and should be avoided in the academic literature.

A second usage suggests that for an animal to suffer it must consciously experience the negative state. It is likely in this sense of the word that some authors interested in the conscious experience of pain in invertebrates and fish use the word suffering not to comment on the quality or magnitude of pain but to emphasize that the pain is, in some way, felt by the animal. For example, Chandroo *et al.* (2004, p. 241) wrote: 'Affective states of pain, fear, and psychological stress are likely to be experienced by fish. This implies that like other vertebrates, fish have the capacity to suffer.' Other authors (e.g. Sherwin, 2001) are less explicit, but comment on cognitive complexity in the animals of interest, suggesting that some level of cognitive complexity is associated with the ability consciously to perceive pain or other types of negative affect.

I suggest that this usage is also weak, at least as applied to most species studied by animal welfare scientists. Unless specified otherwise, research on affect in the animal welfare literature assumes that the experience is felt and matters to the animal in the sense that the animal would work to access or avoid conditions that result in positive or negative affective states respectively. The exact nature of subjective experiences in any other individual is ultimately unknowable, but any applied welfare research (e.g. the search for effective analgesics) can only proceed on the presumption that the individual is in some way aware of the affect.

For a more meaningful usage of the term, let us turn to Marian Dawkins's elegant and provocatively titled book, *Animal Suffering* (Dawkins, 1980). Early in the text (p. 25), Dawkins contends that suffering includes 'a wide range of unpleasant emotional states', but also argues that the magnitude and duration of these experiences is important to consider. For example, she argues (p. 76) that:

> 'Not all fear, frustration or conflict indicates suffering. But prolonged or intense occurrences of these same states may indicate great suffering.' Later in the book (p. 114), she adds that if 'animals had shown evidence of a build up of physiological symptoms that were known to be precursors of disease ... we might conclude that the animals were suffering. If, in addition they showed every sign of trying to escape from their cages and avoiding them when given the opportunity, the evidence they were suffering would be even stronger.'

Dawkins also acknowledges the difficulty in establishing a clear line where suffering can be said to begin. For example, on p. 115 she writes: 'There is a subjective element, for example, in deciding how much fear, conflict, etc. constitutes "suffering".'

In these quotes, we see the word 'suffering' used to distinguish severe or prolonged negative affect, from experiences that are short or mild. But does the experience of negative affect necessarily change qualitatively when severe or prolonged? If not, it may be more parsimonious to specify the duration and magnitude directly. Consider, for example, a dairy cow that becomes lame. This may be quantified using a gait score of 4 on a lameness scale that spans from 1 to 5. On closer examination, we may find that the cow has a sole ulcer, a condition that we know persists for several weeks. The cow in all likelihood is experiencing pain, explaining the altered gait, and this pain is likely intense and relatively long lasting. What extra evidence should we require before we call this suffering?

11.4 Literature on Suffering in Human Patients

As we have seen, other authors have defined suffering as experiencing a negative subjective state. Some go on to specify that the animal must be aware of this state, and others specify that the affect must be of considerable magnitude or duration. I suggest that these conditions are necessary but not sufficient, as can be seen from the literature on suffering in human patients. Work by social scientists has attempted to understand the effects of pain and illness on patients, primarily by recording the patients' own narratives and using these stories to understand how people distinguish pain from suffering (see Bendelow, 2006). For example, Black (2007, p. 37) recounts the story of 'Mrs S' – a patient diagnosed with severe arthritis. When asked if she was suffering, she said no, explaining: 'I do not let myself. I force myself to face my fears. I do not allow myself to feel sorry for myself. I look at some of the young people here in the building [where Mrs S is living] who have never walked a step in their life. I can walk and dance.'

A common theme from these studies is that the way patients respond to pain relates to a range of other factors affecting their mental and emotional state. When patients experience multiple, interacting negative states, they become more likely to characterize their state as suffering. One obvious example is sadness or depression. Here, we can turn to a story told by Mrs S of when, as a young woman, she was ill with a sexually transmitted disease acquired from her husband. Although she experienced 'severe pain' as a result of the illness, she did not 'suffer' while she remained hopeful that she would be able to reconcile with her husband from whom she had become estranged. Paradoxically, the suffering 'didn't start until after I was well', when Mrs S learned that her husband had moved to another city and filed for divorce (Black, 2007, p. 40).

Fear is another powerful concomitant. Cassell (1982) recounts that one patient required 'small doses of codeine' for pain when she thought that this pain was due to sciatica, but required much higher doses when she found out that the cause was cancer. This and other examples indicate that when pain is associated with fear, the likelihood of suffering increases; fear that the pain will increase to the extent that it can no longer be controlled with analgesics, fear that it will last forever, fear that the patient will become overwhelmed by the pain, or fear that the pain is a sign of a serious disease.

Frank recounts from his own experience how fear can turn even the perception of disease into suffering, while the lack of fear can make intense pain manageable:

> During the month between getting the bad news of the irregular chest X ray and receiving the good news about the biopsy, my paradoxical condition was to enjoy very good health in the verified presence of serious illness. I experienced the suffering of illness without experiencing any disease. My bizarre confluence of circumstances turned that month into a controlled experiment in pure suffering. I contrast that experience to a recent attack of tendonitis in my shoulder. The tendonitis caused extraordinary pain ... but I knew what was happening and had reasonable assurance that the acute phase would not last long. The pain lasted a couple of months, especially at night, but once I knew there was no damage requiring surgery, the pain was nothing more than pain. I had no particular plans that involved more than the minimal use of my shoulder. So here is the reverse experiment: pain with more annoyance than suffering.
>
> (Frank, 2001, p. 354)

Fear associated with loss of control is a recurrent theme in studies of human suffering. As described by Cassell (1999, p. 531): 'Suffering can start with anguish over the possibility that if the symptom continues, the patient will be overwhelmed or lose control—"I won't be able to take it".' Loss of control is sometimes characterized as loss of the essence of who you consider yourself to be as a person. In this way, suffering is seen as 'a state of severe distress associated with events that threaten the intactness of the person' (Cassell, 1982, p. 640) and 'defined as a specific state of distress that occurs when the intactness or integrity of the person is threatened or disrupted'. As succinctly expressed by Frank (2001, p. 355), 'to suffer is to lose your grip'. The importance of loss of 'personhood' to the concept of suffering can be seen even in the definition of torture as 'the use of methods upon a person intended to obliterate the personality of the victim or to diminish his physical or mental capacities, even if they do not cause physical pain' (Organization of American States, 1995).

One way to operationalize such threats to personhood is to see when patients are no longer able to do those things that are most important to them. Snyder (2005, p. 69) makes this point when he states that: 'Pain is a physical sensation of discomfort, whereas suffering taps the degree to which a person has let the pain prevent him or her from doing the important things in life.' This view was also illustrated in the above quote by Mrs S, who stated that her current arthritis was not a source of suffering as she was still able to 'walk and dance'.

A reduction in the rate of previously highly motivated behaviour may also be indicative of anhedonia, when patients no longer take pleasure in what they previously enjoyed. Anhedonia is a sign of clinical depression; thus, reduced rates of these behaviours may be both a cause of suffering (when patients can no longer perform the activity, threatening their personhood) and a sign of suffering (if sufficiently depressed to show signs of anhedonia).

How patients understand the meaning of their pain can be another predictor of suffering. In one sense, this can again be related to fear, as in the cancer example discussed earlier, but patients also report positive meanings to their pain, such as that associated with childbirth, and even that which is self-inflicted in attempts at 'spiritual cleansing' (Cassell, 1982). Thus, pain that is perceived to have value is less likely to be associated with suffering by the patient.

11.5 How Can This Be Applied to Animals?

11.5.1 'Asking' the animals

The insights into suffering from human patients described above are all based on verbal reports, suggesting that these ideas will be difficult to apply to the animals we are most interested in. But older readers will remember a time when verbal reports were thought to be the only legitimate way of assessing pain; the astounding development of innovative behavioural methods of pain assessment in recent years, including work on facial expressions of pain in infants and laboratory rodents (e.g. Langford *et al.*, 2010), illustrates that creative (and reliable) methods can be developed and used to 'ask' non-verbal humans and other animals a range of questions about their affective states. The challenge now is to be able to assess in animals the combination of factors that, in human patients, are known to contribute to suffering.

11.5.2 Reduced performance of motivated behaviours

Of the various contributing factors discussed in the previous section, the easiest to apply to animals is a decline in (previously) motivated behaviours. The animal behaviour literature provides a wide range of methods to record the frequency and duration of behaviours, and even to assess changes in the animal's motivation to express these behaviours (Fraser and Nicol, 2011). Thus, my first suggestion for identifying animal suffering is that the animal is experiencing a negative affect such as pain *and* the animal shows a reduced frequency (or magnitude) of behaviours that are known to be important to it. I also suggest distinguishing between those behaviours that decline as a direct result of the pain (i.e. because performing the behaviour is now painful) and those behaviours that reduce in frequency as a result of changes in mood indicative of depression. The former may be associated with suffering in some cases (as in the example of Mrs S, who would likely suffer if she was no longer able to walk or dance) but not in others (as in the example of Frank, who was unconcerned that his injured shoulder may have prevented certain activities). The latter (i.e. pain accompanied by evidence of depression) provides stronger evidence of suffering.

The examples above were all of positively motivated behaviours – those that animals do because they find them pleasurable or otherwise rewarding. But the idea of reduced performance of motivated behaviours can also be applied to those that are negatively motivated – those that animals perform to avoid or escape stimuli that they find aversive. In cases where animals have repeatedly experienced aversive conditions that they cannot avoid, such as repeated exposure to electric shock, animals can develop a condition called 'learned helplessness', characterized by extreme lethargy and reduced attempts at avoiding the shock. Learned helplessness in humans is associated with frustration and depression (Eisenstein *et al.*, 1997). Animals experiencing pain and showing signs of learned helplessness should thus be considered to be suffering.

11.5.3 Indicators of sadness or depression

People in sad moods rate their probability of success less than do happy individuals in tasks involving some risk, so performance in tasks involving risk might also be used to assess mood in animals (Paul *et al.*, 2005). A series of experiments has also shown evidence of such 'cognitive bias' in animals (Mendl *et al.*, 2009). In one experiment, researchers trained rats to distinguish between positive and negative training tones. Some rats were kept in standard housing and others were kept in unpredictable housing conditions known to produce depression-like responses in rats. Rats from the unpredictable housing were more likely to show negative responses to ambiguous test tones (at frequencies intermediate to the two training tones).

Responses to ambiguous stimuli provide just one way of assessing negative or positive hedonic states in animals, but these recent successes show that such assessments are possible. These advances provide another immediate method by which we could begin the assessment of animal suffering – for example, by combining measures of cognitive bias with measures of pain. According to this paradigm, the pain can be labelled as 'suffering' if it is accompanied by low mood, as evidenced by a negative bias in response to ambiguous stimuli.

Heather Neave, a member of our research group at the University of British Columbia, Canada, recently applied this approach in examining calf responses in the hours after hot-iron disbudding, a procedure that is known to cause post-operative pain for 24 h or more (Stafford and Mellor, 2005). Neave trained calves to distinguish between video screens that were either white or red, and then tested calves with intermediate colours (i.e. shades of pink). Before disbudding, calves showed the expected generalization response, with intermediate responses to the intermediate colours, but after disbudding (when they were experiencing post-operative pain) calves showed a negative response bias, responding less often than expected to the intermediates but with no change in responding to the positive and negative training stimuli (Neave *et al.*, 2013). I conclude that the calves were suffering; there was evidence that they experienced both pain and low mood associated with the pain during the hours after dehorning.

This way of thinking also suggests that the treatment of suffering may require more than just treating the underlying negative affective state such as pain; it will also require treating the low mood or, better still, addressing the conditions that resulted in the low mood. In some cases, the low mood will be a direct result of the affective state (as evidenced by the change in bias with the onset of pain in the disbudding example above), in which case treating the pain alone may be adequate.

11.5.4 Indicators of fear or anxiety

Fear and anxiety in animals have been defined by Boissy (1995, p. 166) as 'emotional states that are induced by the perception of any actual danger (fear state) or potential danger (anxiety state) that threatens the well-being of the individual, and which are characterized as a feeling of insecurity'. Fear responses can include alarm calling, escape, defence reactions and, in some cases, immobility. Many domestic animals may perceive their human caretakers as predators, eliciting some of these responses (Rushen *et al.*, 1999). These findings suggest that many domestic animals may be experiencing fear, and in situations with poor facilities or handling, this fear may be long lasting or severe. Suffering is more likely if pain or other negative affect is associated with fear, for example, fear associated with inappropriate handling methods that cause pain. This conclusion suggests that research on negative handling and negative human–animal relationships (especially those that cause both fear and pain) deserves special focus if we are to confront animal suffering.

The fear of losing control is frequently referenced in the human literature and therefore also deserves consideration. Thus, providing animals with more opportunity for control may reduce the risk of suffering; providing animals the ability for self-control in painful situations may be especially beneficial. For example, training animals using positive reinforcement to approach a handler who will perform a routine veterinary procedure, such as taking a blood sample, may be less likely to result in suffering than if the same blood sample was taken from the animal using physical restraint with no opportunity for the animal to control the duration or severity of the procedure. Another approach is to provide opportunities for self-medication with analgesics, anxiolytics, etc. (e.g. Sherwin and Olsson, 2004), allowing the animals to mitigate the negative feelings of, say, pain and also to learn that the mitigation methods are under their own control.

11.5.5 Loss of 'personhood'

The concept of personhood, let alone the loss thereof, is difficult to apply to animals, but given the importance of this idea to the concept of suffering in the human literature, some attempt to understand what this means for animals is required.

Consider *phajaan* – the traditional process of 'breaking' wild elephants. The young calf is separated from its mother, restrained in a crush or shackled by the legs, and goaded by the mahout and others, who use an ankus (a pointed goad) or other tools to inflict pain. The idea is to break the will of the elephant, making it much easier to work with and handle once the process is complete (Kontogeorgopoulos, 2009). In this case, fear and pain are used in combination with the explicit intention of removing the will of the animal to act independently. Phajaan would seem to be an explicit attempt to remove the 'personhood' of the young elephant and provides a clear example of suffering. Phajaan would also seem to meet the definition of torture as outlined above.

11.6 Next Steps

A series of scientific approaches has been reviewed above. Although these approaches may be challenging to implement, they are well within the grasp of modern animal welfare science. We can measure motivation and changes in motivation and recognize the signs of learned helplessness; we have scientific methods of assessing mood, including techniques such as cognitive bias testing; and we can recognize signs of fear and anxiety in animals, and even identify assaults to an animal's 'personhood'. This provides a research agenda for the coming decade and beyond for scholars who wish to address the issue of suffering seriously. Challenges remain in finding ways of addressing other factors that are thought to be important in cases of human suffering. For example, Cassell suggests that assessments of suffering require a qualitative assessment of the patient's condition that goes beyond the physical evidence we have concentrated on to date.

> Suffering is related to the severity of the affliction, but that severity is measured in the patient's terms and is expressed in the distress they are experiencing, their assessment of the seriousness or threat of their problem and how impaired they feel themselves to be. The language that describes and defines the patient's suffering is different from the language of medicine – there is too often an actual disconnect between the case history and the patient's narrative. Herein lies one of the reasons for the inadequate relief of suffering. Physicians are trained primarily to find out what is wrong with the body ... in terms of diseases or pathophysiology: they do not examine what is wrong with the person. It would seem, from looking at training programmes and physicians' actions, that people, with all their ideas, conceptions and misconceptions, fears and fancies, and misleading behaviours, are too often seen as something a physician has to get out of the way in order to diagnose and treat diseases and their manifestations. When physicians attend to the body rather than to the person, they fail to diagnose suffering.
>
> (Cassell, 1999, p. 532)

Researchers working on suffering in human patients have developed more quantitative instruments that, with some imagination, may be applicable to animals. For example, Bruera *et al.* (1991) developed the 'Edmonton Symptom Assessment System', a rating scale used to evaluate signs of fatigue, depression, anxiety and pain. Others have focused on other factors contributing to suffering, including loneliness, frustration and feelings of uselessness (Wilson and Cleary, 1995). One survey (Vodermaier *et al.*, 2009, p. 1482) reported 1416 studies testing the efficacy of questionnaires in assessing emotional distress in cancer patients. According to the authors, among the most useful scales were those that assessed domains such as 'depressive symptoms, anxiety symptoms, quality of life (global), quality of life (number of days impaired), perceived social support, and social support desired'. Mount *et al.* suggest a number of strategies for improving quality of life in human patients:

> Identify sources of healing connections for this person before and during illness. Minimize uncertainty: for example, introduce caregivers by name and occupation; discuss hospital routines, assessment, treatment options, related plans, possible side effects, and anticipated timing of intervention ... Promote a calming, pleasant atmosphere characterized by efficiency, accompaniment, and caring, thus promoting a sense of security.
>
> (Mount *et al.*, 2007, p. 386)

With some imagination, much of this could be applied to animals, and there has been some interest in the animal welfare literature on how assessments of quality of life could be performed (e.g. FAWC, 2009). Much of the advice listed above (consistency of routine, familiar comfortable conditions, known and trusted caretakers, calming atmosphere, etc.) could be used in good husbandry manuals today, but the role of these practices in reducing the opportunity for suffering suggest that these should be applied especially for ill animals and those experiencing pain and distress.

The literature on human patients underscores the importance of social support, especially as a means of buffering the negative effects of stressors (Cohen and Wills, 1985), and the effects of social buffering have been seen in animal studies examining the effects of access to a social partner before, during and after exposure to a stressor (Hennessy *et al.*, 2009). Providing animals with a social partner is thus likely to help mitigate negative affective responses to procedures that are unpleasant for the animals; for example, dairy calves vocalize much less in response to weaning from milk if they are kept in the company of a familiar pen mate (De Paula Vieira *et al.*, 2010). Moreover, isolating animals from familiar social companions, especially those that are closely bonded, may be especially likely to contribute to suffering when animals are subjected to pain, fear and other negative affects. In some cases, domestic animals may form social bonds with human caregivers, and in these cases, the human may be a source of social support.

The nature and strength of the relationships with caretakers not only influences the likelihood that negative affect will lead to suffering; it also influences the likelihood that this suffering will be recognized and treated. As Charmaz (1999, p. 375) points out, people vary the moral significance they attribute to suffering depending on their relationship with that individual: 'A person's moral standing also reflects prior relationships and the web of reciprocities within them. Hence, moral claims of suffering wither when relationships are strained and reciprocities have waned.'

Thus, animals that are considered to be of low value, like a chronically lame dairy cow no longer producing much milk and unable to become pregnant, may be doubly damned as both more likely to experience suffering and less likely to receive compassionate treatment from her caregiver. This reasoning suggests that work to build relationships between animals and their caretakers is important to reducing suffering in animals.

11.7 The Last Word

It seems appropriate to leave the last word to E.J. Cassell, whose 1982 paper inspired many of the studies cited above. Cassell wrote as a physician who treated human patients, but if we substitute just a few words (such as 'caregiver' for 'physician' and 'animal' for 'person'), we find a call to arms for addressing suffering in animals (and see Fig. 11.1).

> A distinction based on clinical observations is made between suffering and physical distress. Suffering is experienced by persons, not merely by bodies, and has its source in challenges that threaten the intactness of the person as a complex social and psychological entity. Suffering can include physical pain but is by no means limited to it. The relief of suffering and the cure of disease must be seen as twin obligations of a medical profession that is truly dedicated to the care of the sick. Physicians' failure to understand the nature of suffering can result in medical intervention that (though technically adequate) not only fails to relieve suffering but becomes a source of suffering itself.
>
> (Cassell, 1982, p. 641)

11.8 Conclusions

- The existing literature in animal welfare science uses the term 'suffering' in three ways: as an embellishment when we describe negative affect in animals, to imply conscious experience of negative affect and to identify negative affect that is severe or prolonged. All three uses are weak and should be avoided.
- Human patients most commonly characterize their condition as suffering when negative affective states are combined or interact, especially with fear. For example, suffering may be likely when repeated poor handling subjects the animal to pain and the animal learns to fear the handler.
- The literature on human patients also points to indicators of suffering that may be applicable to other animals. These include reduced performance of motivated behaviours, learned helplessness and loss of 'personhood'.

Questions for discussion

1. Can you think of situations in your own life when you have felt pain but when you would not say that you were suffering? Can you also think of situations when you have been suffering? How do the two kinds of situations differ?

Fig. 11.1. Robbie (pictured above playing in the sand at the local beach) is a 9-year old boxer with degenerative myelopathy, resulting in the loss of muscle control to his hindquarters. His caregivers have gone to great lengths to avoid his suffering. For example, a cart that supports his hindquarters (pictured in the background) allows Robbie to go on valued walks and socialize with other dogs, and may also provide him with a greater sense of control. Photo credit: Leanne McConnachie.

2. Suffering can mean different things to different authors. Should animal welfare science avoid considering this type of subjective issue and focus instead on issues like disease and longevity?
3. Suffering requires that pain or other negative affective states are felt by and matter to the animal. What evidence should be required to justify these assumptions?
4. If phajaan (the process of 'breaking' wild elephants) provides a clear example of suffering, what about methods commonly used to 'break' other animals such as horses?

Acknowledgements

I am grateful to Kevin Brazil for introducing me to the literature on suffering in human patients. I thank Leanne McConnachie for the photograph and story of Robbie, and Mike Appleby, Marian Dawkins, David Fraser, Maria Hötzel, Jeff Rushen, Peter Sandøe and Marina von Keyserlingk for comments on earlier drafts of the manuscript.

References

Bendelow, G.A. (2006) Pain, suffering and risk. *Health, Risk and Society* 8, 59–70.

Black, H.K. (2007) Is pain suffering? A case study. *International Journal of Aging and Human Development* 64, 33–45.

Boissy, A. (1995) Fear and fearfulness in animals. *Quarterly Review of Biology* 70, 165–191.

Bruera, E., Kuehn, N., Miller, M.J., Selmser, P. and MacMillan, K. (1991) The Edmonton Symptom Assessment System (ESAS): a simple method for the assessment of palliative care patients. *Journal of Palliative Care* 7, 6–9.

Cassell, E.J. (1982) The nature of suffering and the goals of medicine. *New England Journal of Medicine* 306, 639–645.

Cassell, E.J. (1999) Diagnosing suffering: a perspective. *Annals of Internal Medicine* 131, 531–534.

Chandroo, K.P., Duncan, I.J.H. and Moccia, R.D. (2004) Can fish suffer?: perspectives on sentience, pain, fear and stress. *Applied Animal Behaviour Science* 86, 225–250.

Charmaz, K. (1999) Stories of suffering: subjective tales and research narratives. *Qualitative Health Research* 9, 362–382.

Cohen, S. and Wills, T.A. (1985) Stress, social support, and the buffering hypothesis. *Psychological Bulletin* 98, 310–357.

Dawkins, M.S. (1980) *Animal Suffering: The Science of Animal Welfare*. Chapman and Hall, New York.

De Paula Vieira, A., von Keyserlingk, M.A.G. and Weary, D.M. (2010) Effects of pair versus single housing on performance and behavior of dairy calves before and after weaning from milk. *Journal of Dairy Science* 93, 3079–3085.

EFSA (European Food Safety Authority) (2012) Guidance on risk assessment for animal welfare. *EFSA Journal* 10, 2513, 30pp.

Eisenstein, E.M., Carlson, A.D. and Harris, J.T. (1997) A ganglionic model of 'learned helplessness'. *Integrative Physiological and Behavioral Science* 32, 265–271.

Elwood, R.W. (2011) Pain and suffering in invertebrates? *ILAR Journal* 52, 175–184.

FAWC (Farm Animal Welfare Council) (2009) *Farm Animal Welfare in Great Britain: Past, Present and Future*. Farm Animal Welfare Council, London.

Frank, A.W. (2001) Can we research suffering? *Qualitative Health Research* 11, 353–362.

Fraser, D. and Nicol, C.J. (2011) Preference and motivation research. In: Appleby, M.C., Mench, J.A., Olsson, I.A.S. and Hughes, B.O. (eds) *Animal Welfare*, 2nd edn. CAB International, Wallingford, UK, pp. 183–199.

Hennessy, M.B., Kaiser, S. and Sachser, N. (2009) Social buffering of the stress response: diversity, mechanisms, and functions. *Frontiers in Neuroendocrinology* 30, 470–482.

Kontogeorgopoulos, N. (2009) The role of tourism in elephant welfare in northern Thailand. *Journal of Tourism* 10, 1–19.

Langford, D.J., Bailey, A.L., Chanda, M.L., Clark, S.E., Drummond, T.E., Echols, S., *et al.* (2010) Coding of facial expressions of pain in the laboratory mouse. *Nature Methods* 7, 447–449.

Mendl, M., Burman, O.H.P., Parker, R.M.A. and Paul, E.S. (2009) Cognitive bias as an indicator of animal emotion and welfare: emerging evidence and underlying mechanisms. *Applied Animal Behaviour Science* 111, 161–181.

Mount, B.M., Boston, P.H. and Cohen, S.R. (2007) Healing connections: on moving from suffering to a sense of well-being. *Journal of Pain and Symptom Management* 33, 372–388.

Neave, H.W., Daros, R.R., Costa, J.H.C., von Keyserlingk, M.A.G. and Weary, D.M. (2013) Pain and pessimism: dairy calves exhibit negative judgment bias following hot-iron disbudding. PLoS ONE 8, e80556.

Organization of American States (1995) Inter-American Convention to Prevent and Punish Torture. *OAS Treaty Series*, 67.

Paul, E.S., Harding, E.J. and Mendl, M. (2005) Measuring emotional processes in animals: the utility of a cognitive approach. *Neuroscience and Biobehavioral Review* 29, 469–491.

Rushen, J., Taylor, A.A. and de Passillé, A.M. (1999) Domestic animals' fear of humans and its effect on their welfare. *Applied Animal Behaviour Science* 65, 285–303.

Sherwin, C.M. (2001) Can invertebrates suffer? Or, how robust is argument-by-analogy? *Animal Welfare* 10, 103–118.

Sherwin, C.M. and Olsson, I.A.S. (2004) Housing conditions affect self-administration of anxiolytic by laboratory mice. *Animal Welfare* 13, 33–38.

Snyder, C.R. (2005) Physical pain does not equal suffering: a personal commentary on coping. *Journal of Loss and Trauma* 10, 51–71.

Stafford, K.J. and Mellor, D.J. (2005) Dehorning and disbudding distress and its alleviation in calves. *Veterinary Journal* 169, 337–349.

Vodermaier, A., Linden, W. and Siu, C. (2009) Screening for emotional distress in cancer patients: a systematic review of assessment instruments. *Journal of the National Cancer Institute* 101, 1464–1488.

Wilson, I.B. and Cleary, P.D. (1995) Linking clinical variables with health-related quality of life. A conceptual model of patient outcomes. *Journal of the American Medical Association* 273, 59–65.

Index